Environmental Biotechnology

Environmental Biotechnology

Monika Jain

Alpha Science International Ltd.
Oxford, U.K.

Environmental Biotechnology
292 pgs. | 65 figs. | 04 tbls.

Monika Jain
Department of Biotechnology
Sharda University
Greater Noida, UP

ALPHA SCIENCE INTERNATIONAL LTD.
7200 The Quorum, Oxford Business Park North
Garsington Road, Oxford OX4 2JZ, U.K.

www.alphasci.com

Printed from the camera-ready copy provided by the Author.

ISBN 978-1-84265-814-7

Printed in India

Acknowledgement

First I want to thank God for giving me strength for completing this book. It is an honor for me to thank those who made this book possible. I owe my deepest gratitude to my loving husband, Devendra, whose encouragement, guidance and support from the initial to final level enabled me to develop this book. I would also like to thank my parents, Mr. Rajendra Kumar Jain and Smt. Sharda Jain, who always encouraged me to write this book. I would also like to show my gratitude to my brother, Rohit, who always inspired me to complete this work. I would also like to thank my friends and my colleagues who helped me in number of ways.

Lastly I offer my regards to all of those who supported me in any respect during the completion of the book.

Monika Jain

Preface

Biotechnological processes to protect the environment have been used for almost a century now, even longer than the term 'biotechnology' exists. Municipal sewage treatment plants and filters to purify town gas were developed around the turn of the century. Environmental biotechnology is the application of all components of biotechnology to solve environmental problems. The prime target of this science is the abatement of pollution through bioremediation/ biotreatment or supporting as resources for human use in non polluting ways.

The development of modern biotechnology has been accompanied by the establishment or adaptation of regulations to deal with genetically modified organisms. On the whole it encompasses aspects of natural resources management, the treatment of waste and control of pollution. Thus the major areas of understanding are environmental pollution abatement through biodegradation, biotransformation, bioaccumulation of toxicity like organics, metals, oil and hydrocarbons, dyes, detergents etc. It also covers energy management through production of non-conventional and non-polluting energy sources like biodiesel, methanol, biogas, bio-hydrogen etc. Agricultural application of biofertilizers, biopesticides, recovery of resources from toxic or nontoxic wastes through biotechnological approach, use of biosensors for pollution monitoring and several other allied issues are important aspects of environmental biotechnology.

Biomining is economically sound hydrometallurgical process with lesser environmental problem than conventional commercial application. However, it is an inter-disciplinary field involving metallurgy, chemical engineering, microbiology and molecular biology. It has tremendous practical application. In a country like India biomining has great national significance where there is vast unexploited mineral potential. Through application of biotechnical methods, enzyme bioreactors are being developed that will pretreat some industrial waste and food waste components and allow their removal through the sewage system rather than through solid waste disposal mechanisms. Waste can also be converted to biofuel to run generators.

The "Environmental Biotechnology" has been written to meet the requirements of the students of B.Sc., M.Sc., B.Tech. and M.Tech. in Biotechnology of various universities. It provides an insight into the basics of the biotechnological processes and methods for pollution control, biodegradation of Xenobiotic compounds, production of Biodegradable plastics, Biomining and Bioleaching. It also covers various methods for solid waste and radioactive waste management.

Though I have tried to make the book student friendly and comprehensive in all respects, it is possible that some errors might still have been left due to oversight. Any suggestions or constructive criticism from the readers will be highly appreciated and considered.

Monika Jain

Contents

1

Environment: Basic Concepts and Issues

Environment means surrounding external conditions influencing development or growth of people, animal or plants; living or working conditions etc. It can be defined variously as:

"Environment is the representative of physical components of the earth where man is an important factor affecting the environment".

"Environment refers to the sum total of condition, which surround point in space and time".

The scope of the term Environment has been changing and widening by the passage of time. In the primitive age, the environment consisted of only physical aspects of the planet earth. Land, air and water as biological communities. As the time passed, man extended his environment through his social, economic and political functions. It is in nature that physical component of the planet earth, i.e. land, air, water etc., support and affect life in the biosphere.

"A person's environment consists of the sum total of the stimulations which he receives from his conception until his death."

It can be concluded from the above definition that environment comprises of various types of forces such as physical, intellectual, economic, political, cultural, social, moral and emotional. Environment is the sum total of all the external forces, influences and conditions, which affect life, nature, behavior, growth, development and maturation of living organisms. The environment consists of four segments as under:

Atmosphere: The atmosphere implies the protective blanket of gases, surrounding the earth. The atmosphere of the Earth serves as a key factor in sustaining the planetary ecosystem. The thin layer of gases that envelops the Earth is held in place by the planet's gravity. Dry air consists of 78% nitrogen, 21% oxygen, 1% argon and other inert gases such as carbon dioxide. The remaining gases are often referred to as trace gases among which are the greenhouse gases such as carbon dioxide, methane, nitrous oxide, ozone and water vapor. Filtered air includes trace amounts of many other chemical compounds. Air also contains a variable amount of water vapor, suspensions of water droplets and ice crystals seen as clouds. Many natural substances may be present in tiny amounts in an unfiltered air sample including dust, pollen and spores, sea spray, volcanic ash and meteoroids. Various industrial pollutants also may be present such as chlorine (elementary or in compounds), fluorine compounds, elemental mercury, and sulphur compounds such as sulphur dioxide.

The ozone layer of the Earth's atmosphere plays an important role in depleting the amount of ultraviolet (UV) radiation that reaches the surface. As DNA is readily damaged by UV light, this serves to protect life at the surface. The atmosphere also retains heat during the night, thereby reducing the daily temperature extremes. Significance of atmosphere:

a) It sustains life on the earth.
b) It saves life from the hostile environment of outer space.
c) It absorbs most of the cosmic rays from outer space and a major portion of the electromagnetic radiation from the sun.
d) It transmits only ultraviolet, visible, near infrared radiation (300 to 2500 nm) and radio waves. (0.14 to 40 m) while filtering out tissue-damaging ultraviolet waves below about 300 nm.

Hydrosphere: The Hydrosphere comprises all types of water resources oceans, seas, lakes, rivers, streams, reservoir, polar ice caps, glaciers, and ground water.

a) 97% of the earth's water supply is in the oceans.
b) About 2% of the water resources are locked in the polar ice caps and glaciers.
c) Only about 1% is available as fresh surface water i.e. rivers, lakes, streams, and ground water fit for human consumption and other uses.

Lithosphere: Lithosphere is the outer mantle of the solid earth. It consists of materials occurring in the earth's crusts and the soil e.g. minerals, organic matter, air and water.

Biosphere: Biosphere indicates the realm of living organisms and their interactions with environment i.e. atmosphere, hydrosphere and lithosphere. The biosphere is the global sum of all ecosystems. It can also be called the zone of life on Earth. A closed (apart from solar and cosmic radiation) and self-regulating system. From the broadest bio-physiological point of view the biosphere is the global ecological system integrating all living beings and their relationships, including their interaction with the elements of the lithosphere, hydrosphere and atmosphere. The biosphere is postulated to have evolved beginning through a process of biogenesis or biopoesis at least some 3.5 billion years ago.

ELEMENTS OF ENVIRONMENT

Environment is constituted by the interacting systems of physical, biological and cultural elements inter-related in various ways, individually as well as collectively. These elements may be explained as under:

(i) Physical elements: Physical elements are as space, landforms, water bodies, climate, soil, rocks and minerals. They determine the variable character of the human habitat, its opportunities as well as limitations.
(ii) Biological elements: Biological elements such as plants, animals, microorganisms and man constitute the biosphere.
(iii) Cultural elements: Cultural elements such as economic, social and political elements are essentially man made features which make cultural milieu.

NATURAL ENVIRONMENT

The natural environment comprises of all naturally occurring surroundings and conditions in which living things grow and interact on Earth. These include complete landscape units that function as natural systems without major human intervention as well as plants, animals, rocks and natural phenomena occurring within their boundaries. They also include non-local or universal natural resources that lack clear-cut boundaries such as air, water and climate. The concept of the natural environment can be distinguished by components:

(i) Complete ecological units that function as natural systems without massive human intervention including all vegetation, microorganisms, soil, rocks, atmosphere and natural phenomena that occur within their boundaries.
(ii) Universal natural resources and physical phenomena that lack clear-cut boundaries such as air, water, climate, energy, radiation, electric charge and magnetism (not originating from human activity).
(iii) The natural environment is contrasted with the built environment which comprises the areas and components that are strongly influenced by humans. A geographical area is regarded as a natural environment.

MAN AND NATURAL ENVIRONMENT

Over the past decade or so, more and more attention is being paid all over the world to man's environment on which human existence depends and the maintenance of which is now increasingly being considered as essential for mankind.

Environment refers to those natural things that surround us and that are essential to sustain human life such as the earth's atmosphere, healthy air and drinkable water together with the non-essentials that help to make life sustainable such as wild animals or wild places or human living space.

With the passage of time mankind is realizing that preserving the essential ingredients of life and the rich natural diversity of the planet is indeed worthwhile. Thus, protecting and saving the environment involves preserving Nature's gifts to mankind as much as possible, and in as good a condition as practicable.

Pollution, especially in the industrialized belt, and the ecological crisis are not wholly new or novel. Though ecology is a comparatively new science, the idea behind it, i.e. the preservation of natural resources, is almost as old as man.

Now the realization has dawned on humanity that we have been destroying valuable resources and that there must indeed be a limit to our plans and ambitions for development, expansion and growth. Murder of the environment which involves senseless poisoning of the earth, air and water, destruction of forest wealth, may be described as "ecocide".

Our rivers are polluted; the Himalayan ecology is seriously endangered in many ways. The consequences of such continuous and reckless use of trees and other natural resources would be disastrous. There are also the dangers from chemical pollution from radio-active wastes and other wastes from homes, factories, hospitals and laboratories and from other foreign matter that keeps entering the atmosphere. It is feared by experts that if the energy of the sun is hindered, if the natural processes of purification and elimination are reversed and if the reckless destruction and pollution continues, mankind may return to the dreaded ice age.

Therefore, preservation and restoration of Nature's balance is vital and efforts are being made for that purpose at both national and international levels.

NATURAL RESOURCE

Natural resources occur naturally within environments that exist relatively undisturbed by mankind in a natural form. A natural resource is often characterized by amounts of biodiversity and geodiversity existent in various ecosystems. Natural resources are derived from the environment. Some of them are essential for our survival while most are used for satisfying our wants. Natural resources may be further classified in different ways.

Natural resources are materials and components (something that can be used) that can be found within the environment. Every man-made product is composed of natural resources (at its fundamental level). A natural resource may exist as a separate entity such as fresh water, air, as well as a living organism such as a fish. Or it may exist in an alternate form which must be processed to obtain the resource such as metal ores, oil and most forms of energy.

There is much debate worldwide over natural resource allocations; this is partly due to increasing scarcity (depletion of resources) but also because the exportation of natural resources is the basis for many economies (particularly for developed nations).

Some Natural resources can be found everywhere such as sunlight and air. When this is so, the resource is known as an ubiquitous (existing or being everywhere) resource. However, most resources are not ubiquitous. They only occur in small sporadic areas. These resources are referred to as localized resources. There are very few resources that are considered inexhaustible (will not run out in foreseeable future). These are solar radiation, geothermal energy, and air (though access to clean air may not be). The vast majority of resources are however exhaustible, which means they have a finite quantity and can be depleted if managed improperly. The natural resources are materials which living organisms can take from nature for sustaining their life or any components of the natural environment that can be utilized by man to promote his welfare.

TYPES OF NATURAL RESOURCES

There are various methods of categorizing natural resources. These include source of origin, stage of development and by their renewability.

On the Basis of Origin

Biotic

Biotic resources are obtained from the biosphere (living and organic material) such as forests, animals, birds, fish and the materials that can be obtained from them. Fossil fuels such as coal and petroleum are also included in this category because they are formed from decayed organic matter.

Abiotic

Abiotic resources are those that come from non-living, non-organic material. Examples of abiotic resources include land, fresh water, air and heavy metals including ores such as gold, iron, copper, silver etc. **(Fig 1.1)**.

On the Basis of Stage of Development

Potential Resources

Potential resources are those that exist in a region and may be used in the future. For example, petroleum may exist in many parts of India, in sedimentary rocks. But until the time it is actually drilled out and put into use it remains a potential resource.

Actual Resources

Actual resources are those that have been surveyed, their quantity and quality determined and are being used in present times. The development of an actual resource such as wood processing depends upon the technology available and the cost involved.

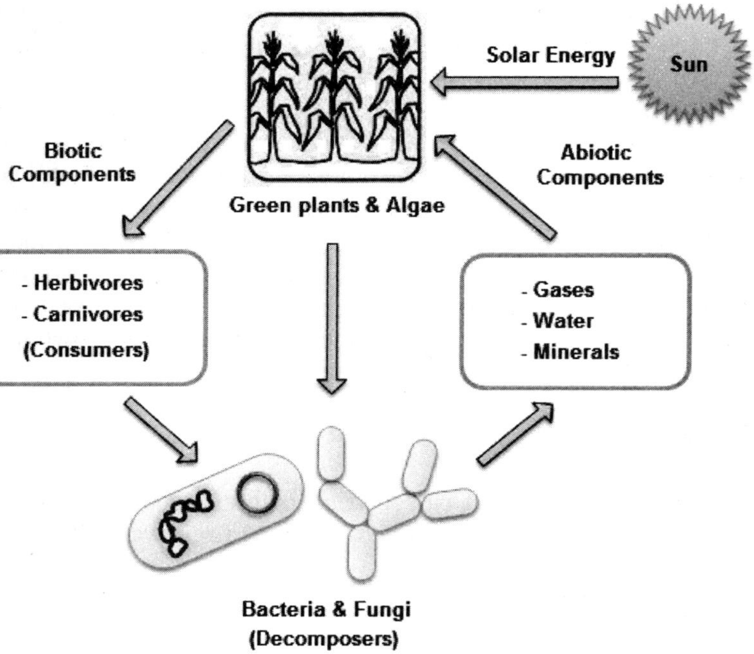

Fig. 1.1: Biotic and Abiotic components

Reserve Resources

The part of an actual resource which can be developed profitably in the future is called a reserve resource.

Stock Resources

Stock resources are those that have been surveyed but cannot be used by organisms due to lack of technology e.g. hydrogen.

On the Basis of Renewability

Renewable Resources

Renewable resources are ones that can be replenished naturally. Some of these resources are sunlight, wind, etc. are continuously available and their quantity is not noticeably affected by human consumption. Though many renewable resources do not have such a rapid recovery rate, these resources are susceptible to depletion by over-use. Resources from a human use perspective are classified as renewable only so long as the rate of replenishment/recovery exceeds that of the rate of consumption.

Non-renewable Resources

Non-renewable resources are resources that form extremely slowly and those that do not naturally form in the environment. Minerals are the most common resource included in this category. By the human perspective resources are non-renewable when their rate of consumption exceeds the rate of replenishment/recovery. A good example of this are fossil fuels which are in this category because their rate of formation is extremely slow (potentially millions of years), meaning they are considered non-renewable. Some resources actually naturally deplete in amount without human interference. The most notable of these being radioactive elements such as uranium which naturally decay into heavy metals. Of these, the metallic minerals can be re-used by recycling them but coal and petroleum cannot be recycled.

CONSERVATION OF NATURAL ENVIRONMENT

The rapid pace of industrialization, the discharge of untreated waste, the effluents, the widespread neglect of environment, the ceaseless pollution through human ignorance and use of open spaces near rivers as public conveniences is resulting in the pollution of major rivers, estuaries, sea-fronts, creeks and water channels.

World has reached a stage when the absorptive and assimilative capacity has been overused or misused, resulting in pollution and environmental degradation. It is true that the natural environment has an enormous capacity to accept and absorb most wastes; what is discarded by one species is often used up by another. But with more industrial centers emerging in many states, chimneys releasing dark smoke and coal-dust into the atmosphere, drains getting choked with chemical and poisonous wastes, the dangers of pollution are increasing month by month.

The environment that we live in and make use of is being stripped off its precious components day by day. There are many angles from which the problem of environmental problem can be studied. Similarly, many different views come in to play if we need to find practical solutions to these problems. There are innumerable factors associated with conservation of environment and also these problems are interrelated. The activities as well as the lifestyle of human beings too play an important role in how the surroundings/environment is affected. Till date it is one of the major factors that have affected the environment of the Earth as it is in constant conflict with it. Let us therefore, understand the different environmental problems of today and try to find the best possible solutions to them.

Challenges

It is the common understanding of natural environment that underlies environmentalism - a broad political, social and philosophical movement that advocates various actions and policies in the interest of protecting what remains in the natural environment, restoring or expanding the role of nature in this environment. While true wilderness is increasingly rare wild nature (e.g., unmanaged forests, uncultivated grasslands, wildlife, wildflowers) can be found in many locations previously inhabited by humans. Goals commonly expressed by environmental scientists include:

(i) Reduction and clean up of pollution, with future goals of zero pollution.
(ii) Cleanly converting non-recyclable materials into energy through direct combustion or after conversion into secondary fuels.
(iii) Reducing societal consumption of non-renewable fuels.
(iv) Development of alternative, green, low-carbon or renewable energy sources.
(v) Conservation and sustainable use of scarce resources such as water, land, and air.
(vi) Protection of representative or unique or pristine ecosystems.
(vii) Preservation of threatened and endangered species' extinction.

The establishment of nature and biosphere reserves under various types of protection and most generally, the protection of biodiversity and ecosystems upon which all human and other life on earth depends. Conservation is the wise use of natural resources (nutrients, minerals, water, plants, animals, etc.) and cultural resources (different groups of people from different parts of the world). It may also include protecting the large collections of resources that make up a habitat or environment. Rapid change can force animals, plants, places, or people to become endangered or extinct. Conservation will allow future generations to enjoy natural resources such as clean rivers and lakes, wilderness areas, a diverse wildlife population, healthy soil, and clean air. Many natural resources are necessary for our survival. It may be difficult to imagine that we could ever run out of fresh water, clean air, and good soil for growing food. But in some places this is already happening.

Most scientists believe that conservation of biodiversity should be the most important thing for people and governments to work on. Preserving biodiversity will help ensure a healthy planet for all living things. Conserving biodiversity makes sure that as many living things survive as possible. This is important to humans because there are still many undiscovered plant and animal species. For example, some plants contain chemicals that may help cure or prevent diseases like cancer. These plants may grow in a place no one has explored. If we destroy a place's biodiversity, we may never know that there was a species of plant that could have provided an important cure.

Different types of environments exist all over our planet. Each type of environment has its own special weather, plants and animals, water supply, minerals, and other resources. Everything is in balance. Life in that environment depends on all its resources for survival. If any of the resources are removed too fast or removed completely, it destroys the balance. When an environment is out of balance many living things may have trouble surviving. Scientists and conservationists believe the best way to preserve cultural and biological diversity is by protecting large pieces of the environment. For example, they may encourage governments to protect an entire wetlands or a large section of tropical rainforest. A second way is to connect environments with corridors where wildlife and plants can move without being detected.

ENVIRONMENTAL IMPACT ASSESSMENT

An environmental impact assessment is an assessment of the possible positive or negative impact that a proposed project may have on the environment, together consisting of the environmental, social and economic aspects.

The purpose of the assessment is to ensure that decision makers consider the ensuing environmental impacts when deciding whether to proceed with a project. The International Association for Impact Assessment (IAIA) defines an environmental impact assessment as "the process of identifying, predicting, evaluating and mitigating the biophysical, social, and other relevant effects of development proposals prior to major decisions being taken and commitments made." EIAs are unique in a sense that they do not require adherence to a predetermined environmental outcome but rather they require decision makers to account for environmental values in their decisions and to justify those decisions in light of detailed environmental studies and public comments on the potential environmental impacts of the proposal.

There are eight guiding principles **(Fig. 1.2)** that govern the entire process of EIA and they are as follows:

Participation: An appropriate and timely access to the process for all interested parties.

Transparency: All assessment decisions and their basis should be open and accessible.

Certainty: The process and timing of the assessment should be agreed in advanced and followed by all participants.

Accountability: The decision-makers are responsible to all parties for their action and decisions under the assessment process.

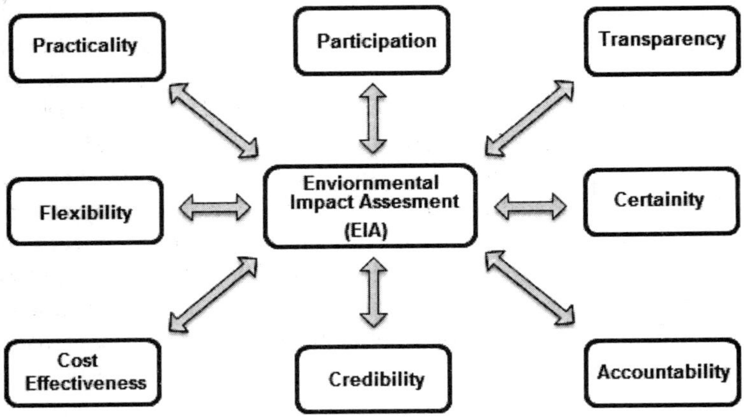

Fig. 1.2: Principles of Environment Impact Assessment

Credibility: Assessment is undertaken with professionalism and objectivity.

Cost-effectiveness: The assessment process and its outcomes will ensure environmental protection at the least cost to the society.

Flexibility: The assessment process should be able to adapt to deal efficiently with any proposal and decision making situation.

Practicality: The information and outputs provided by the assessment process are readily usable in decision making and planning.

Environmental Impact Assessment in India

An EIA concentrate on problems, conflicts and natural resource constraints which might affect the viability of a project. It also predicts how the project could harm to people, their homeland, their livelihoods and the other nearby developmental activities. After predicting potential impacts, the EIA identifies measures to minimize the impacts and suggests ways to improve the project viability. The aim of an EIA is to ensure that potential impacts are identified and addressed at an early stage in the projects planning and design. To achieve this aim, the assessment finding are communicated to all the relevant groups who will make decisions about the proposed projects, the project developers and their investors as well as regulators, planners and the politicians. Having read the conclusions of an environmental impact assessment project planners and engineers can shape the project so that its benefits can be achieved and sustained without causing adverse impacts. In recent years, major projects have encountered serious difficulties because insufficient account has been taken of their relationship with the surrounding environment. Some projects have been found to be unsustainable because of resource depletion. Others have been abandoned because of public opposition, financially encumbered by unforeseen costs, held liable for damages to natural resources and even been the cause of disastrous accidents. Given this experience, it is very risky to undertake finance or approve a major project without first taking in to account its environmental consequences and then designing the project so as to minimize adverse impacts.

In India many of the developmental projects till as recently as the 1980s were implemented with very little or no environmental concerns. The environmental issues began receiving attention when a national committee on environmental planning and coordination was set up under the 4th five year plan (1969-1978). Till 1980, the subjects of environment and forests were the concern of the Dept of Science and Technology and Ministry of Agriculture respectively. Later, the issues were formally attended by the Dept of Environment which was established in 1980. This was then upgraded to the Ministry of Environment & Forest in 1985. In 1980, clearance of large projects from the environmental angle became an administrative requirement to the extent that the planning commission and the central investment board sought proof of such clearance before according financial sanction.

Five year later, the Department of Environment and Forests, Government of India, issued guidelines for Environmental Assessment of river valley projects. These guidelines require various studies such as impacts on forests and wild life in the submergence zone, water logging potential, upstream and downstream aquatic ecosystems and fisheries, water related diseases, climatic changes and seismicity.

A major legislative measures for the purpose of environmental clearance was in 1994 when specific notification was issued under section 3 and rule 5 of the environment protection Act , 1986 called the "Environment impact Assessment Notification 1994". The procedure involves the following steps:

Screening

The screening is the first and simplest tier in project evaluation. Screening helps to clear those types of projects which from past experience are not likely to cause significant environmental problems.

Preliminary Assessment

If screening does not clear a project, the developer may be required to undertake a preliminary Assessment. This involves sufficient research, review of available data and expert advice in order to identify the key impacts of the project on the local environment, predict the extent of the impacts and briefly evaluate their importance to decision makers.

Formation of EIA Team

If after reviewing a preliminary assessment the competent authority deems that a full EIA is needed, the next step for the project developer is the preparation of the EIA report.

Scoping

The first task of the EIA study team is scoping the EIA. The aim of scoping is to ensure that the study address all the issues of importance to the decision makers. First of all the team's outlook is broadened by the discussions (with the project proponents, decision makers, the regulatory agency, scientific institutions, local community representative and others) to include all the possible issues and concerns raises by various groups. Then the study team selects primary impacts for the EIA to focus upon depending on the basis of magnitude, geographical extent, significance to decision makers or because the area is special locally (e.g. soil erosion, the presence of an endangered species, or a nearby historical sites) or is an eco-sensitive area.

Main EIA

After "scoping" the main EIA begins. It identifies the result of the project, predicts the extents of the changes, evaluation of the changes by the project, etc.

Documentation

The last step in the EIA process- In documenting an EIA, this means identifying the key decisions makers, perceiving the question they will be asking and providing them with straight forward answers formatted for easy interpretation in relation to their decision making (e.g. tables, graphs, summary, points). Successful EIA documentation is more readily produced if the audience and their needs are established at the start of the EIA, and then made to affect how the research is focused and reported.

A completed assessment document must then be made available for public inspection. It is sometimes necessary for the agency to put together a follow-up document, commonly known as a supplemental assessment. This may occur if major changes are made in project plans or in the environment that could be affected.

EIA Benefits and Flaws

EIA generates huge benefits in selection of project location, process, design, development actions, and decision-making; however, in the current practice of EIA there are a number of flaws, shortcomings and deficiencies. The table below summarizes apparent benefits and flaws of the EIA (Table 1.1).

One significant factor that could help improve the EIA process is good education and training. Currently, very few educational and training courses exist in developing countries that properly consider various EIA methodologies available in depth. Information on the legal and regulatory frameworks and institutional arrangements are also necessary. Education and training process are important since the fundamental factors behind all EIA predictions are still the best professional judgment and/or experiences with similar projects implemented elsewhere.

Table 1.1 EIA Benefits and Flaws

Benefits	**Flaws**
Provides systematic methods of impact assessment	Time-consuming
Estimates the cost/benefit trade-off of alternative action	Costly
Facilitates the public participation	Little public participation in actual implementation
Provides an effective mechanism for coordination environmental integration negotiations feedback	Unavailability for reliable data (mostly in developing countries)
Top-level decision making	Too focused on scientific analysis (sometimes)
Triggers an institutional building	Poor presentation of EIA report (bulky volumes, scientific explanation, difficult to understand)
Achieve a balance between the impact of developmental and environmental concern	Compliance monitoring after EIA is seldom carried out

2

Environmental Biotechnology

Environmental Biotechnology is the multidisciplinary integration of sciences and engineering in order to utilize the huge biochemical potential of microorganisms, plants and parts thereof for the restoration and preservation of the environment and for the sustainable use of resources. Environmental biotechnology is defined as the development, use and regulation of biological system e.g. cells, cell compartments, enzymes, for remediation of contaminated environments (land, air, water and sediments), and for environment-friendly processes (green manufacturing technologies and sustainable development). The primary role of environmental biotechnology is to develop better approaches for sustainable development and for understanding processes in the natural environment. The driving force of biotechnology is abilities of microorganisms to utilize various carbon sources naturally occurring as pollutants.

Environmental biotechnology can simply be described as "the optimal use of nature, in the form of plants, animals, bacteria, fungi and algae, to produce renewable energy, food and nutrients in a synergistic integrated cycle of profit making processes where the waste of each process becomes the feedstock for another process".

SCOPE OF ENVIRONMENTAL BIOTECHNOLOGY

Biotechnological processes to protect the environment have been used for almost a century now, even longer than the term 'biotechnology' exists. Municipal sewage treatment plants and filters to purify town gas were developed around the turn of the century. They proved very effective although at the time, little was known about the biological principles underlying their function. Since that time our knowledge base has increased enormously. This briefing paper describes the state-of-the-art and possibilities of environmental biotechnology. It also deals with the societal aspects of environmental biotechnology.

Environmental biotechnology is the application of all components of biotechnology to solve environmental problems. The prime target of this science is the abatement of pollution through bioremediation/ biotreatment or supporting as resources for human use in non polluting ways. It can also help in cleaner production of existing products. On the whole it encompasses aspects of natural resources management, the treatment of waste and control of pollution. Thus the major areas of understanding are environmental pollution abatement through biodegradation, biotransformation, bioaccumulation of toxicity like organics, metals, oil & hydrocarbons, dyes, detergents etc. It also covers energy management through production of non-conventional and non-polluting energy sources like biodiesel, methanol, biogas, bio-hydrogen etc. Agricultural application of biofertilizers, biopesticides, Recovery of resources from toxic or nontoxic wastes through biotechnological approach, use of biosensors for pollution monitoring and several other allied issues are also included in it **(Fig 2.1)**.

Biotechnological techniques to treat waste before or after it has been brought into the environment are described and exemplified in the section on bioremediation. Biotechnology can also be used to develop products and processes that generate less waste and use less non-renewable resources and energy. In this respect biotechnology is well positioned to contribute to the development of a more sustainable society, a principle which was advocated in the Brundtland Report in 1987 and in Agenda 21 of the second Earth Summit in Rio de Janeiro in 1992 and which has been widely accepted in the mean time. Recombinant DNA technology has improved the possibilities for the prevention of pollution and holds a promise for a further development of bioremediation. The development of modern biotechnology has been accompanied by the establishment or adaptation of regulations to deal with genetically modified organisms.

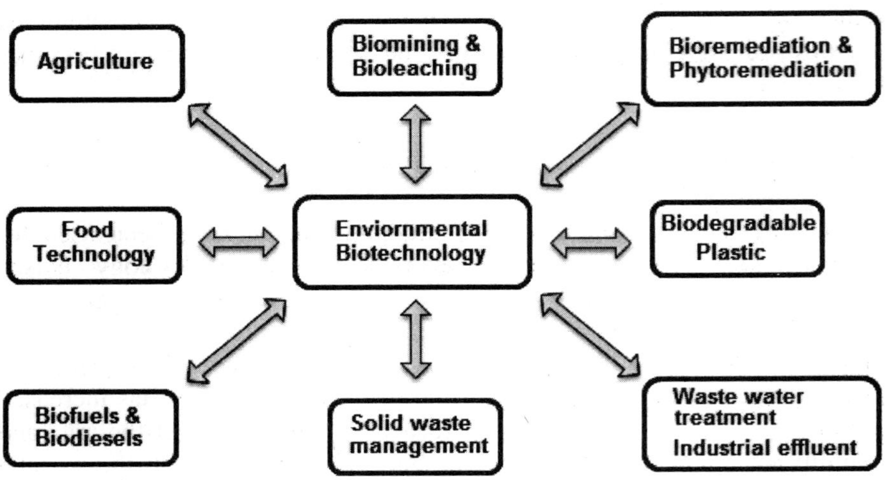

Fig. 2.1: Scope of Environmental Biotechnology

BIOTECHNOLOGY AND POLLUTION CONTROL

Any product, by product or residue that cannot be used profitably is called a waste. A waste product is regarded as a pollutant when it damages the environment. Often wastes and pollutants are intricately linked. In simple words, pollutants are generally wastes, but all wastes are not pollutants. Wastes may be:

(i) Biological

(ii) Chemical

(iii) Physical in nature and may originate from the following activities:

a) Manufacturing

b) Agriculture and dairy

c) Energy production

d) Transport

e) House building and housekeeping

Rapid industrialization, urbanization and other developments have resulted in a threatened clean environment and depleted natural resources. The World Summit on Sustainable Development (WSSD) held in Johannesburg (S. Africa) during August 26 to September 4, 2002 has assessed the global changes which occurred after 1992 i.e. the Rio Earth Summit. The two problems which are receiving constant attention of environmentalists are:

(i) Control of environmental pollution

(ii) Conservation of nature and natural resources

Biotechnology is being considered as an emerging technology in environmental protection. It involves the use of genetically manipulated microorganisms which are likely to prove more suitable for pollution control due to their versatility and adaptability to changing environments. Sulphur oxide, nitrous oxides, volatile organic compounds and particulates are the four major components of air pollution and are responsible for environmental hazards. These can be easily removed from the environment by using the biotechnological methods of their disposal.

Environmental Biotechnology applies scientific and engineering principles to the design of manufacturing and combustion processes to reduce air pollutant emissions to acceptable levels. Scrubbers, electrostatic precipitators, catalytic converters and various other processes are utilized to remove particulate matter, nitrogen oxides, sulfur oxides, volatile organic compounds (VOC), reactive organic gases (ROG) and other air pollutants from flue gases and other sources prior to allowing their emission to the atmosphere.

Scientists have developed air pollution dispersion models to evaluate the concentration of a pollutant at a receptor or the impact on overall air quality from vehicle exhausts and industrial flue gas stack emissions. To some extent, this field overlaps the desire to decrease carbon dioxide and other greenhouse gas emissions from combustion processes.

Three types of biological waste gas purification systems are in operation. These are:

(i) Bioscrubbers

(ii) Biofilters

(iii) Biotrickling filters

Waste Water Management

Most urban and many rural areas no longer discharge human waste directly to the land through outhouse, septic and/or honey bucket systems, but rather deposit such waste into water and convey it from households via sewer systems. Engineers and scientists develop collection and treatment systems to carry this waste material away from where people live and produce the waste and discharge it into the environment. In developed countries, substantial resources are applied to the treatment and detoxification of this waste before it is discharged into a river, lake or ocean system. Developing nations are striving to obtain the resources to develop such systems so that they can improve water quality in their surface waters and reduce the risk of water-borne infectious disease.

There are numerous wastewater treatment technologies. A wastewater treatment train can consist of a primary clarifier system to remove solid and floating materials, a secondary treatment system consisting of an aeration basin followed by flocculation and sedimentation or an activated sludge system and a secondary clarifier, a tertiary biological nitrogen removal system, and a final disinfection process. The aeration basin/activated sludge system removes organic material by growing bacteria (activated sludge). The secondary clarifier removes the activated sludge from the water. The tertiary system, although not always included due to costs, is becoming more prevalent to remove nitrogen and phosphorus and to disinfect the water before discharge to a surface water stream or ocean outfall.

Engineers and scientists work to secure water supplies for potable and agricultural use. They evaluate the water balance within a watershed and determine the available water supply, the water needed for various needs in that watershed, the seasonal cycles of water movement through the watershed and they develop systems to store, treat, and convey water for various uses. Water is treated to achieve water quality objectives for the end uses. In the case of potable water supply, water is treated to minimize the risk of infectious disease transmission, the risk of non-infectious illness, and to create a palatable water flavor. Water distribution systems are designed and built to provide adequate water pressure and flow rates to meet various end-user needs such as domestic use, fire suppression and irrigation.

Wastewater is typically categorized into one of the following groups:

(i) Municipal wastewater (domestic wastewater mixed with effluents from commercial and industrial works, can be pre-treated or non pre-treated).

(ii) Commercial and industrial wastewater (pre-treated or not pre-treated).

(iii) Agricultural wastewaters.

The effluent components may be of chemical, physical or biological in nature and they can induce an environmental impact, which includes changes in aquatic habitats and species structure as well as in biodiversity and water quality.

The three major groups of biological processes: aerobic, anaerobic, combination or in sequence to offer greater levels of treatment. The main objectives of wastewater treatment processes can be summarized as:

(i) Reduction of biodegradable organics content (BOD_5).

(ii) Reduction/removal of recalcitrant organics.

(iii) Removal of heavy/toxic metals.

(iv) Removal/reduction of compounds containing P and N (nutrients).

(v) Removal and inactivation of pathogenic microorganisms and parasites.

Treatment Biotechnology for Raw Water

Raw water drawn from various sources is not fit enough for human consumption directly without subjecting it to treatment. Certain gases, traces of minerals and other undesirable substances get dissolved in the raw water as a result of various hydrological processes. During runoff on the earth surface, raw water picks up soil, garbage, sewage, pesticides and other physical, chemical and bacteriological agents including human and animal wastes.

The following treatment methods are generally applicable each suing different technologies:

a) Screening

b) Sedimentation

c) Filtration

d) Disinfection

e) Softening

f) Desalination

g) Demineralization

The technologies available for treatment of municipal water supplies, for industry or for large communities are the following:

a) Clarification

b) Filtration

c) Ultra-filtration

d) Flocculation

e) Reverse-osmosis

f) Electro dialysis

g) Water-softening

h) Fluoride Removal

i) Disinfection

j) Iron removal

The available technologies and equipment are suitable and adequate for rural sectors and at small communities' level. R&D agencies such as NEERI Nagpur etc. have done considerable work in this field. The technology gap is in respect to more sophisticated techniques such a reverse osmosis and ultra filtration including membrane technology and surface water pollution.

Biotechnology in Solid Waste Management

Wherever men live, the waste follows. Since he is responsible for waste, it is his pious duty to manage waste in scientific way. A contaminated environment places people at obvious risk of exposure to pathogens that lead to infection and diseases. In urban areas, waste generation and its handling has become an important issue. Many biotechnologies developed for waste recycling in agriculture were suitably modified to manage urban waste as well. To begin with, a household level solid waste management system was developed. Subsequently, scientific waste management systems for residential colonies were also developed. The biotechnology for waste water recycling process is based on treatment during organic and hydraulic loading. Suitable biological filters can be designed, made with natural materials. Microbes inoculated in these filters can absorb both organic and inorganic impurities of waste water. The system efficiency can be further improved by utilizing selected plants, where root zone treatment can be carried out. These systems remove all harmful microbes and reduce residual BOD and COD.

Type of Waste Materials

Basically the waste can be categorized into (a) Biodegradable Waste and (b) Non Biodegradable Waste. Biodegradable waste or, if recovered, the biodegradable component of mixed wastes (e.g. paper & food waste) can be composted or anaerobically digested to produce soil improvers and renewable fuels. Non-biodegradable waste is that waste which cannot be degraded.

Waste is broadly segregated into solid, liquid and gaseous waste materials. Waste falls into a number of different waste types. It can exist in any phase of matter (solid, liquid, or gas) or as waste heat. When released in the latter two states the wastes can be referred to as emissions. It is usually strongly linked to pollution.

Some components of waste can be recycled once recovered from the waste stream, e.g. plastic bottles, metals, glass or paper. There are many different waste types or waste streams which are produced by a variety of processes. Each waste type has different methods of associated waste management. The following is a list of waste types:

a) Animal byproducts.

b) Biodegradable waste.

c) Bulky waste.

d) Clinical waste.

e) Construction and demolition waste (C&D waste).

f) Domestic waste.

g) Electronic waste (E-waste).

h) Farm waste.

i) Food waste and Household waste etc.

Waste Management Tools

The management of different waste requires different kind of procedures to handle as the different toxic compounds that might be present in one may not be present in the other. Mother Nature recycles all types of waste materials. We are producing more waste materials then nature is capable of recycling and thus it results into pollution.

A very attractive way to change garbage into rich humus is to utilize the services of earthworms. Vermiculture means farming of earthworms through bio-degradable material. Earthworms are nature's fertilizer factory.

Physically they are crushers and grinders, due to action of their gizzard. There are thousands of different species of worms, but the best manure worms is *Eisenia fetida*, as it works everywhere, in the indoor as well as at outdoor. They are a surface dwelling variety of worms that hate the light and reproduce at an amazing rate. The aim of this management is:

a) Creating awareness at household level regarding the issue of garbage and its proper management.

b) Collection of wastes in segregated form.

c) Conversions of organic/wet waste into high quality Vermicompost.

d) Appropriate use of Vermicompost for planting trees, gardens and lawns etc. to make clean and green environment surrounding our houses.

Scientific waste management in urban areas, waste including human excreta and waste from polluting industries are disposed through sewers. This pollutes the environment, underground water and exposes people to infection. Open decomposition of solid waste and sewer water through existing river systems takes very long period in natural treatment while causing many health hazards. In many countries, initiatives to develop the skills through recycled waste materials for producing vegetables by the poor have brought about excellent results. In some Latin American countries, vegetable production in urban areas through waste recycling have not only been able to reduce the direct and indirect costs associated with waste disposal but it has also simultaneously been able to solve the problems of urban sanitation, while becoming an income generating activity as well.

Solid Waste Management can be done by:

(i) Segregation of wastes at the time of generation.

(ii) Collection of segregated waste.

(iii) Recycle waste materials.

(iv) Process non recyclable materials in decentralized manner.

The various solid waste management scientific tools used are as follows:

***Physical Treatment*:**

(i) Reduction in volume and weight.

(ii) Reducing the size of waste material.

Biological Treatment:

(i) Solid sanitizers.

(ii) Deodorizers.

(iii) Microbial cultures for accelerated decomposition.

(iv) Microbes for waste conversion.

The biological processes improving fast are shown among the future technologies. In these processes the biological materials are used as degraders, which process raw wastes to remove the contaminants in them. Biotechnological processes are used for wastewater treatment, gas treatment and disposal of solid wastes. Also, these processes can be utilized for the production of biogas and hydrogen as new energy resources. For preventing environmental pollution in environmental engineering, activated sludge process, trickling filters, biotrickling filters, oxidation ponds, anaerobic treatment, composting units and biogas reactors are used extensively among the waste treatment technologies.

Biofuels Production

Biofuels are produced from living organisms or from metabolic by-products (organic or food waste products). In order to be considered a Biofuels the fuel must contain over 80 percent renewable materials. It is originally derived from the photosynthesis process and can therefore often be referred to as a solar energy source. There are many pros and cons to using Biofuels as an energy source. A biofuel is a type of fuel whose energy is derived from biological carbon fixation. Biofuels include fuels derived from biomass conversion, as well as solid biomass, liquid fuels and various biogases. Although fossil fuels have their origin in ancient carbon fixation, they are not considered biofuels because they contain carbon that has been "out" of the carbon cycle for a very long time. Biofuels are gaining increased public and scientific attention, driven by factors such as oil price hikes, the need for increased energy security, concern over greenhouse gas emissions from fossil fuels and support from government subsidies.

Bioethanol is an alcohol made by fermentation, mostly from carbohydrates produced in sugar or starch crops such as corn or sugarcane. Cellulosic biomass, derived from non-food sources such as trees and grasses, is also being developed as a feedstock for ethanol production. Ethanol can be used as a fuel for vehicles in its pure form, but it is usually used as a gasoline additive to increase octane no. and improve vehicle emissions. Bioethanol is widely used in the USA and in Brazil.

Biodiesel is made from vegetable oils and animal fats. Biodiesel can be used as a fuel for vehicles in its pure form, but it is usually used as a diesel additive to reduce levels of particulates, carbon monoxide, and hydrocarbons from diesel-powered vehicles. Biodiesel is produced from oils or fats using transesterification and is the most common biofuel in Europe.

In 2010 worldwide biofuel production reached 105 billion liters (28 billion gallons US), up 17% from 2009, and biofuels provided 2.7% of the world's fuels for road transport, a contribution largely made up of ethanol and biodiesel. Global ethanol fuel production reached 86 billion liters (23 billion gallons US) in 2010, with the United States and Brazil as the world's top producers, accounting together for 90% of global production. The world's largest biodiesel producer is the European Union, accounting for 53% of all biodiesel production in 2010. As of 2011, mandates for blending biofuels exist in 31 countries at the national level and in 29 states/provinces. According to the International Energy Agency, biofuels have the potential to meet more than a quarter of world demand for transportation fuels by 2050.

Biodegradable Plastics

Bioplastics represent a relatively new class of materials which have much in common with conventional plastics. What differentiates them is the use of renewable resources in their manufacture, the biodegradability and compostability of many bioplastics products. Their development follows nature's example: 100 billion tonnes of biomass are annually produced from plants, using sunlight and photosynthesis. The same amount biodegrades back into the source materials, carbon dioxide (CO_2) and water, together with small amounts of biomass and minerals. This occurs primarily through biological degradation via numerous microbes. The bioplastics industry's aim is to imitate this closed loop, as it represents the means by which CO_2 emissions can be reduced and fossil resources conserved for future generations.

Bioplastics are not a single class of polymers but rather a family of products which can vary considerably one from the other. A generally recognized definition of the concept does not exist. European Bioplastics, like other associations, regards bioplastics as having two differentiated classes.

Plastics based on renewable resources and biodegradable polymers which meet all criteria of scientifically recognized norms for biodegradability and compostability of plastics and plastic products. In both classes, a high percentage of renewable resources is used in the polymer production. Whereas products from the first group do not necessarily have to be biodegradable or compostable, those from the second group do not necessarily have to be based on renewable materials in order to meet the criteria. Even a number of petrochemical-based polymers are certified biodegradable and compostable. They broaden the range of applications and are often responsible for creating the pre-requisites to enable renewable resources to be used in plastics production.

Plastic Types:

a) Starch-based plastics.

b) Cellulose-based plastics.

c) Some aliphatic polyester.

The aliphatic biopolyesters are mainly polyhydroxyalkanoates (PHAs) like the poly-3-hydroxybutyrate (PHB), polyhydroxyvalerate (PHV) and polyhydroxyhexanoate (PHH).

Applications

Biodegradable bioplastics are used for disposable items, such as packaging and catering items (crockery, cutlery, pots, bowls, and straws). Biodegradable bioplastics are also often used for organic waste bags, where they can be composted together with the food or green waste and decompose in aerobic and anaerobic environments. Some trays and containers for fruit, vegetables, eggs and meat, bottles for soft drinks and dairy products and blister foils for fruit and vegetables are manufactured from bioplastics.

Nondisposable applications include mobile phone casings, carpet fibers, and car interiors, fuel line and plastic pipe applications and new electro active bioplastics are being developed that can be used to carry electrical current. In these areas, the goal is not biodegradability, but to create items from sustainable resources.

Biopesticides

So far, use of synthetic chemical pesticides had been the widely used approach for reducing the estimated 45% gross crop loss due to pests and diseases, amounting to around Rs 290 billion per annum. More and more quantities of chemicals are used for agricultural intensification to feed an ever growing population. In fact, the pest induced loss is on the rise despite increasing usage of pesticides. Fortunately, realization of the negative effects of these chemicals on nature and natural resources like pollution, pesticide residue, pesticide resistance etc; have forced many to shift focus on to more reliable, sustainable and environment friendly agents of pest control, the biopesticides. In spite of the claimed efficacy, their use, however, has remained very low due to a number of socio-economic, technological and institutional constraints. Nonetheless, rise in income levels due to a growing economy coupled with increasing awareness of health related effects of chemical pesticides has increased the demand of organic food. In view of this demand and the government's efforts to mitigate climate change, biopesticides are going to play an important role in future pest management programmes.

Scope of Biopesticides

A pesticide that is of biological origin i.e., viruses, bacteria, pheromones, plant or animal compounds is known as biopesticide. Or simply origin of the active ingredient of a biopesticide is natural not synthetic. They are highly specific affecting only the targeted pest or closely related pests and do not harm humans or beneficial organisms while chemical pesticides are broad spectrum and known to affect non-target organisms including predators and parasites as well as humans.

The striking feature of biopesticides is environment friendliness and easy biodegradability, thereby resulting in lower pesticide residues and largely avoiding pollution problems associated with chemical pesticides. Further, use of biopesticides as a component of Integrated Pest Management (IPM) programs can greatly decrease the use of conventional (chemical) pesticides, while achieving almost the same level of crop yield. However, effective use of biopesticides demands understanding of a great deal about managing pests especially by the end users. In terms of production and commercialization also biopesticides have an edge over chemical pesticides like low research expenditure, faster rate of product development as well as flexible registration process. Biopesticides fall into three major classes.

Microbial pesticides consist of bacteria, entomopathogenic fungi or viruses (and sometimes includes the metabolites that bacteria or fungi produce). Entomopathogenic nematodes are also often classed as microbial pesticides, even though they are multi-cellular.

Plant-incorporated protectants (PIPs) have genetic material from other species incorporated into their genetic material (i.e. GM crops). Biochemical pesticides are naturally occurring substances that control pests by nontoxic mechanisms

Advantages

(i) Do not leave harmful residues.

(ii) Substantially reduced impact on non-target species.

(iii) Can be cheaper than chemical pesticides when locally produced.

(iv) Can be more effective than chemical pesticides in the long-term.

Disadvantages

(i) High specificity, which will require an exact identification of the pest/pathogen and may require multiple pesticides to be used.

(ii) Often slow speed of action, thus making them unsuitable if a pest outbreak is an immediate threat to a crop.

(iii) Often variable efficacy due to the influences of various biotic and abiotic factors (since biopesticides are usually living organisms, which bring about pest/pathogen control by multiplying within the target insect pest/pathogen).

(iv) Living organisms evolve and increase their resistance to biological, chemical, physical or any other form of control. If the target population is not exterminated or rendered incapable of reproduction, the surviving population can acquire a tolerance of whatever pressures are brought to bear, resulting in an evolutionary arms race.

Biodegradation of Xenobiotic Compounds

Use of pesticides has benefited the modern society by improving the quantity and quality of food production. Gradually, pesticide usage has become an integral part of modern agriculture system. Many of the artificially made complex compounds i.e. xenobiotics persist in environment and do not undergo biological transformation. Microorganisms play an important role in degradation of xenobiotics, and maintaining of steady state concentrations of chemicals in the environment. The complete degradation of a pesticide molecule to its inorganic components that can be eventually used in an oxidative cycle removes its potential toxicity from the environment. However, there are two objectives in relation to biodegradation of xenobiotics:

a) Estimation of biodegradation activity evolved and transferred among the members of soil microflora.

b) Device bioremediation methods for removing or detoxifying high concentration of dangerous pesticide residues.

Xenobiotic compounds are human made chemicals that are present in the environment at unnaturally high concentrations. The xenobiotic compounds are either not produced naturally, or are produced at much lower concentrations.

Microorganism has the capability of degrading all naturally occurring compounds; this is known as the principle of microbial infallibility proposed by Alexander in 1965. Microorganisms are also able to degrade many of the xenobiotic compounds, but they are unable to degrade many others.

The compounds that resist biodegradation and thereby persist in the environment are called recalcitrant. The xenobiotic compounds may be recalcitrant due to one or more of the following reasons:

a) They are not recognized as substrate by the existing degradative enzymes.

b) They are highly stable, i.e., chemically and biologically inert due to the presence of substitution groups like halogens, nitro-, sulphonate, amino-, methoxy- and carbamyl groups.

c) They are insoluble in water, or are adsorbed to external matrices like soil.

d) They are highly toxic or give rise to toxic products due to microbial activity.

e) Their large molecular size prevents entry into microbial cells.

f) Inability of the compounds to induce the synthesis of degrading enzymes.

g) Lack of the permease needed for their transport into the microbial cells.

Bioleaching

Nowadays bioleaching occupies an increasingly important place among the available mining technologies. Today bioleaching is no longer a promising technology but an actual economical alternative for treating specific mineral ores. An important number of the current large-scale bioleaching operations are located in developing countries. This situation is determined by the fact that several developing countries have significant mineral reserves and by the characteristics of bioleaching that makes this technique especially suitable for these countries because of its simplicity and low capital cost requirement.

The current situation of commercial-size bioleaching operations and ongoing projects in developing countries is presented and discussed with especial reference to copper and gold mining. It is concluded that this technology can significantly contribute to the economic and social development of these countries.

Microbial leaching is the process by which metals are dissolved from ore bearing rocks using microorganisms. For the last 10 centuries, microorganisms have assisted in the recovery of copper dissolved in drainage from water. Thus biomining has emerged as an important branch of biotechnology in recent years. Microbial technology renders helps in case of recovery of ores which cannot be economically processed with chemical methods, because they contain low grade metals. Therefore, large quantity of low grade ores are produced during the separation of high grade ores. The low grade ores are discarded in waste heaps which enter in the environment. The low grade ores contain significant amount of nickel, lead, and zinc ores which could be processed by microbial leaching. Bioleaching of uranium and copper has been widely commercialized. But large scale leaching process may cause environmental problems when dump is not managed properly.

This result in seepage of leach fluids containing large quantity of metals and low pH into nearby natural water supplies and ground water. Thus, biomining is economically sound hydrometallurgical process with lesser environmental problem than conventional commercial application. However, it is an inter-disciplinary field involving metallurgy, chemical engineering, microbiology and molecular biology. It has tremendous practical application. In a country like India biomining has great national significance where there is vast unexploited mineral potential.

Bioleaching is a method used for extraction of precious and base metals from hard to treat ore with the aid of bacterial microorganism. Bioleaching is used to recover copper, zinc, lead, arsenic, antimony, nickel, molybdenum, gold, silver and cobalt. Bioleaching is considered to be an efficient and an ecological friendly process commonly used by the miners as an alternative method to roasting or smelting, especially when there are lower concentrations of metal in the ore. Bioleaching produces no offensive gases as it is a hydro metallurgical form of treatment. The process involves the bacteria feeding on the nutrients in the ore there by separating the metal. The metal can then be collected from the bottom of the solution. Bioleaching is possible because of the unique microorganism's ability to react and breakdown the mineral deposits in the ore.

FUTURE OF ENVIRONMENTAL BIOTECHNOLOGY

Environmental biotechnology is the used in waste treatment and pollution prevention. Environmental biotechnology can more efficiently clean up many wastes than conventional methods and greatly reduce our dependence on methods for land-based disposal.

Every organism ingests nutrients to live and produces by-products as a result. Different organisms need different types of nutrients. Some bacteria thrive on the chemical components of waste products. Environmental engineers use bioremediation, the broadest application of environmental biotechnology, in two basic ways. They introduce nutrients to stimulate the activity of bacteria already present in the soil at a waste site, or add new bacteria to the soil. The bacteria digest the waste at the site and turn it into harmless byproducts. After the bacteria consume the waste materials, they die off or return to their normal population levels in the environment.

Bioremediation, is an area of increasing interest. Through application of biotechnical methods, enzyme bioreactors are being developed that will pretreat some industrial waste and food waste components and allow their removal through the sewage system rather than through solid waste disposal mechanisms. Waste can also be converted to biofuel to run generators. Microbes can be induced to produce enzymes needed to convert plant and vegetable materials into building blocks for biodegradable plastics.

In some cases, the byproducts of the pollution-fighting microorganisms are themselves useful. For example, methane can be derived from a form of bacteria that degrades sulfur liquor, a waste product of paper manufacturing. This methane can then be used as a fuel or in other industrial processes.

3

Air Pollution and its Management through Biotechnology

Air is a precious resource that most of us take for granted. Air supplies us with oxygen, which is essential for our bodies to live. Without it, we would die within minutes. Pure air is a mixture of several gases that are invisible and odorless. It consists of about 78% nitrogen, 21% oxygen, and less than 1% of argon, carbon dioxide, and other gases — as well as varying amounts of water vapor. Adults breathe in about 10-20 cubic meters of air every day. That's about 20,000 breaths. Children breathe almost twice that amount because they are smaller, and their respiratory systems are still maturing.

AIR POLLUTION

Air pollution is the introduction of chemicals, particulate matter, or biological materials that cause harm or discomfort to humans or other living organisms, or cause damage to the natural environment or built environment, into the atmosphere. The air pollution is also known as the atmospheric pollution. The W.H.O defined it as the presence of materials in the air which are harmful to the living beings when they cross their threshold concentration levels. The foreign bodies, gases etc. act as an air pollutant. These chemicals are emitted from the refineries, paper mills, ceramics, fertilizers, clay and glass manufacturing industries. The important pollutants involved are the fluorides, sulphur dioxide and hydrogen sulphide.

It occurs most often in cities, where it manifests itself as smog, as well as gases that are not visible to the human eye. In fact, urban air quality and pollution found indoors are the two main pollution problems that the world faces today. Due to the fact that the atmosphere is made of a delicate and complex balance of natural gases that support all living things and the ecosystems of the planet, the addition of pollution can lead to devastating effects in the long term.

There are essentially two types of air pollutants: particles and noxious gases.

Particles are usually released by burning fuel and, in some cases, wood or charcoal. These particles are typically the result of automobiles and homes utilizing fuel as an energy source.

Noxious gases, however, are the source of acid rain and smog. These gases consist of carbons dioxide, sulfur dioxide, and chemical vapors, amongst other things. Often, these are emitted by older, unregulated vehicles and factories. These two harmful types of pollutants can also be found in our homes and places of work, as smoking, fumes from certain hair care and beauty products, and cooking can release these pollutants, as well.

Causes of Air Pollution

There are several main causes of air pollution; the vast majority of them can be attributed to man. Air pollution is the introduction of chemicals, particulates, and biological matter that cause harm to humans, other living organisms, or cause damage to the natural environment into the atmosphere. Stratospheric ozone depletion (contributed to air pollution) has long been recognized as a threat to human health as well as to the Earth's ecosystems. The Earth is capable of cleaning itself of a certain level of pollution, but man-made pollutant have become too numerous for the Earth's natural mechanisms to remove. Some of the sources of air pollution are as follows:

Industrial Emissions

One of the main causes of air pollution is manufacturing. This source of pollution spews particulate matter and chemicals into the atmosphere. The exhaust from a factory includes, sulfur oxides, nitrogen oxides, carbon monoxide and dioxide, as well as volatile organic compounds and particulates. There is not an area of the Earth's atmosphere or an ecosystem that has not been altered by the long term effects of the pollution created by manufacturing. In addition to these typical industrial emission contributors, there are also a number of other culprits. Petroleum refineries release a variety of particulates and hydrocarbons into the Earth's atmosphere. Also, though a bit rarer, radioactive fallout from nuclear facilities is also one of the causes of air pollution.

Fossil Fuels

The burning of fossil fuels is a part of the everyday life of every human on the planet. Fossil fuels are emitted from our cars, and have become a big part of our everyday lives. Fossil fuel is even used to extract greater amounts of fossil fuel from the earth, where it is naturally occurring, and is also used to break it down into its individual components. As a result, this is typically among the highest on every pollution list today. It is also the most controversial of the causes of air pollution, as fossil fuels are a necessary element of today's society, despite the fact that their effects are causing great harm to us and to the environment. Without using fossil fuels, factories would not produce goods and we would not experience the ease of travel that we do today. Essentially, most of us are extremely dependent upon fossil fuels, such as coal and oil. Until an alternative is widely accepted, we run the risk of further damaging the environment and our overall health.

Chemicals

This cause encompasses a number of chemicals, such as those commonly used household cleaning products or painting supplies. However, it can be argued that the main chemicals that most associate with pollution are pesticides. Crop dusting, fumigating homes, and over the counter insect/pest killers emits harmful chemicals into the environment, to contribute to the overall poor air quality that most regions experience today. Another pollutant found in this category is fertilizer dust, which is released into the air from agricultural grade fertilizers.

The Earth itself

Earth contributes to air pollution through volcanic eruptions, wildfires, and other natural process, but it is capable of recovering from those sources. As humans continue to industrialize with a dependence on fossil fuels, we continue to put our planet's future in jeopardy and shorten the longevity of our species. There are a number of air pollutants which come directly from the earth itself, which seems a bit ironic. Methane emitted by animals simply digesting their food or carbon monoxide from wild fires are all contributing factors to the world's pollution problem. On warmer days, there are even some regions where the vegetation can produce great amounts of VOCs, and volcanic activity has the capability to disperse sulfur and chlorine into the air we breathe. Also, the radioactive decay occurring within the Earth's crust emits radon. Radon is a colorless, odorless gas that is the bi-product of radium. It can then collect in buildings and tight spaces that have little or ventilation, causing a radon builds up, which is the second leading cause of lung cancer, smoking being the first. Surprisingly, even dust found in areas with minimal vegetation is considered a source of air pollution.

Pollutants

A substance in the air that can cause harm to humans and the environment is known as an air pollutant. Pollutants can be in the form of solid particles, liquid droplets, or gases. In addition, they may be natural or man-made. Air pollutants can be visible (e.g., the brownish-yellow color of smog) or invisible. Besides affecting human health and the environment, air pollutants can also hamper our ability to see very far (visibility). Air pollution can have local and regional impacts — such as ground-level ozone and wood smoke. It can also have wide-reaching, global effects — such as climate change and depletion of the ozone layer.

Pollutants can be classified as primary or secondary. Usually, primary pollutants are directly emitted from a process, such as ash from a volcanic eruption, the carbon monoxide gas from a motor vehicle exhaust or sulfur dioxide released from factories. Secondary pollutants are not emitted directly. Rather, they form in the air when primary pollutants react or interact. An important example of a secondary pollutant is ground level ozone — one of the many secondary pollutants that make up photochemical smog. Some pollutants may be both primary and secondary: that is, they are both emitted directly and formed from other primary pollutants **(Fig 3.1)**.

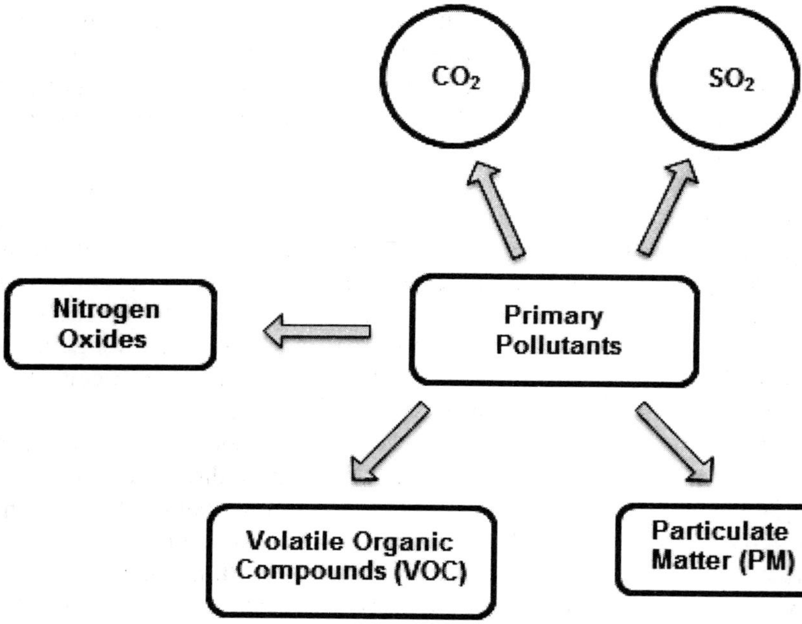

Fig. 3.1: Types of primary pollutants

Primary Pollutants

Primary air pollutants are emitted directly into the air from sources. They can have effects both directly and as precursors of secondary air pollutants (chemicals formed through reactions in the atmosphere), which are discussed in the following section.

Sulfur dioxide (SO_2)

It is a gas formed when sulfur is exposed to oxygen at high temperatures during fossil fuel combustion, oil refining, or metal smelting. SO_2 is toxic at high concentrations, but its principal air pollution effects are associated with the formation of acid rain and aerosols. SO_2 dissolves in cloud droplets and oxidizes to form sulfuric acid (H_2SO_4), which can fall to Earth as acid rain or snow or form sulfate aerosol particles in the atmosphere.

Nitrogen oxides (NO and NO_2, referred together as NO_x)
They are highly reactive gases formed when oxygen and nitrogen react at high temperatures during combustion or lightning strikes. Nitrogen present in fuel can also be emitted as NOx during combustion. Emissions are dominated by fossil fuel combustion at northern mid-latitudes and by biomass burning in the tropics. In the atmosphere NOx reacts with volatile organic compounds (VOCs) and carbon monoxide to produce ground-level ozone through a complicated chain reaction mechanism. It is eventually oxidized to nitric acid (HNO3). Like sulfuric acid, nitric acid contributes to acid deposition and to aerosol formation.

Carbon monoxide (CO)
It is an odorless, colorless gas formed by incomplete combustion of carbon in fuel. The main source is motor vehicle exhaust, along with industrial processes and biomass burning. Carbon monoxide binds to hemoglobin in red blood cells, reducing their ability to transport and release oxygen throughout the body. Low exposures can aggravate cardiac ailments, while high exposures cause central nervous system impairment or death. It also plays a role in the generation of ground-level ozone.

Volatile organic compounds
VOCs including hydrocarbons (C_xH_y) but also other organic chemicals are emitted from a very wide range of sources, including fossil fuel combustion, industrial activities and natural emissions from vegetation and fires. Some anthropogenic VOCs such as benzene are known carcinogens. VOCs are also of interest as chemical precursors of ground-level ozone and aerosols. The importance of VOCs as precursors depends on their chemical structure and atmospheric lifetime, which can vary considerably from compound to compound. Large VOCs oxidize in the atmosphere to produce nonvolatile chemicals that condense to form aerosols. Short-lived VOCs interact with NO_x to produce high ground-level ozone in polluted environments. Methane (CH_4), the simplest and most long-lived VOC, is of importance both as a greenhouse gas and as a source of background troposphere ozone. Major anthropogenic sources of methane include natural gas production and use, coal mining, livestock, and rice paddies.

Secondary Pollutants

Particulate matter
Particulates alternatively referred to as particulate matter (PM) or fine particles, are tiny particles of solid or liquid suspended in a gas. In contrast, aerosol refers to particles and the gas together. Sources of particulate matter can be manmade or natural. Some particulates occur naturally, originating from volcanoes, dust storms, forest and grassland fires, vegetation and sea spray. Human activities, such as the burning of fossil fuels in vehicles, power plants and various industrial processes also generate significant amounts of

aerosols. Averaged over the globe, anthropogenic aerosols—those made by human activities—currently account for about 10 percent of the total amount of aerosols in our atmosphere. Increased levels of fine particles in the air are linked to health hazards such as heart disease, altered lung function and lung cancer. Persistent free radicals connected to airborne fine particles could cause cardiopulmonary disease. Toxic metals, such as lead, cadmium and copper are also responsible for pollution. Chlorofluorocarbons (CFCs) are harmful to the ozone layer, are emitted from products currently banned from use. Particulate matter refers to tiny solid or liquid particles that float in the air. Some particles are large or dark enough to be seen as smoke, soot or dust. Others are so small that they can only be detected with a powerful, electron microscope. PM occurs in two forms: primary and secondary.

Primary PM is emitted directly into the atmosphere by wood burning and fossil fuel burning. Primary PM also includes pollen, spores and road dust. Secondary PM is formed in the atmosphere through chemical reactions involving nitrogen dioxide, sulphur dioxide, volatile organic compounds and ammonia. Particulate matter formed from gaseous primary pollutants and compounds is termed as Photochemical smog. Smog is a kind of air pollution. Classic smog results from large amounts of coal burning in an area caused by a mixture of smoke and sulfur dioxide. Modern smog does not usually come from coal but from vehicular and industrial emissions that are acted on in the atmosphere by ultraviolet light from the sun to form secondary pollutants that also combine with the primary emissions to form photochemical smog.

Ammonia

Ammonia (NH_3) is emitted from agricultural processes. It is normally encountered as a gas with a characteristic pungent odor. Ammonia contributes significantly to the nutritional needs of terrestrial organisms by serving as a precursor to foodstuffs and fertilizers. Ammonia, either directly or indirectly, is also a building block for the synthesis of many pharmaceuticals. Although in wide use, ammonia is both caustic and hazardous. Odors such as from garbage, sewage and industrial processes also act as pollutants. Radioactive pollutants which are produced by nuclear explosions, nuclear events, war explosives and natural processes such as the radioactive decay of radon.

Ground level ozone

Ozone is found in two regions of the Earth's atmosphere – at ground level and in the upper regions of the atmosphere. Both types of ozone have the same chemical composition (O_3). While upper atmospheric ozone protects the earth from the sun's harmful rays, ground level ozone is the main component of smog.

Tropospheric or ground level ozone, is not emitted directly into the air, but is created by chemical reactions between oxides of nitrogen (NO_x) and volatile organic compounds (VOC). Ozone is likely to reach unhealthy levels on hot sunny days in urban environments. Ozone can also be transported to long distances by wind. For this reason, even rural areas can experience high ozone levels. Ground level ozone, which we breathe, can harm our health. Even relatively low levels of ozone can cause health effects. People with lung disease, children, older adults, and people who are active outdoors may be particularly sensitive to ozone. Children are at greatest risk from exposure to ozone because their lungs are still developing and they are more likely to be active outdoors when ozone levels are high, which increases their exposure. Children are also more likely than adults to have asthma.

Peroxyacetyl nitrate (PAN)

It is similarly formed from NOx and VOCs. It is a secondary pollutant present in photochemical smog. It is thermally unstable and decomposes into peroxyethanoyl radicals and nitrogen dioxide gas. It is a lachrymatory substance. Peroxyacetyl nitrate, or PAN, is an oxidant more stable than ozone. Hence, it is better capable of long-range transport than ozone. It serves as a carrier for oxides of nitrogen (NOx) into rural regions and causes ozone formation in the global troposphere. The formation of PAN on a secondary scale becomes an issue when ethanol is used as an automotive fuel. Acetaldehyde emissions increase, which subsequently react in the atmosphere to form smog.

Effects of Air pollution

A variety of air pollutants have known or suspected harmful effects on human health and the environment. Pollutants from the sources may not only prove a problem in the immediate vicinity of these sources but can travel long distances. The various effects are as follows:

Biological Systems

On animals

Air pollution is a major alarming environmental problem affecting the developing and developed countries adversely. The health effects caused by air pollutants may range from biochemical and physiological changes to difficulty in breathing, wheezing, coughing. All pollutants that are inhaled have serious impact on human health affecting the lungs and the respiratory system. Individual reactions to air pollutants depend on the type of pollutant, the degree of exposure, the individual's health status and genetics. These pollutants are also deposited on soil, plants and in the water, further contributing to human exposure. The most harmful of the gases and pollutants that emitted are particulate matter, carbon dioxide, polycyclic organic matter, and formaldehyde.

On plants

Photochemical smog has a deleterious effect on plants. Chronic exposure to ozone may weaken plants and make them susceptible to disease and reduce crop yield. Acids (acid rain) enhances the uptake of toxic heavy metals by the plants.

On non-biological Systems

Climatic effects

Green house effect

In the environment, carbon dioxide is confined exclusively in troposphere. In dense concentration, it can act as serious pollutant. The temperature of the earth surface has been maintained by the energy of the sun's rays that strike the planet and heat that gets radiated back into the space. Some of the rays that penetrate carbon dioxide layer are able to strike earth and get converted into heat. The heated earth is able to reradiate this absorbed energy as radiations of longer wavelength. At higher CO_2, concentrations, much of heat gets absorbed by the CO_2. And water in the atmosphere adds to the heat that has been already present. Thus the earth's atmosphere heats up. This phenomenon of heating of earth's atmosphere is termed as green house effect.

Global warming

Since Chlorofluorocarbons and NO deplete ozone layer, they allow the penetration of harmful UV radiation to strike the earth's surface. CFCs can hang in the atmosphere for decades. They slowly diffuse into the stratosphere, where UV-radiation decomposes them into atomic chlorine that triggers a lengthy $Cl\ O_x$ chain reaction forming various oxides. This chain eats away the ozone. This led UV rays to enter into the atmosphere. This warms up the troposphere to an uncomfortable level and raise surface temperature causes global warming.

Acid rain

Acid rain means the presence of excessive acids in rain water. Sulphur dioxide (SO_2) and nitrogen dioxide (NO_2) are highly soluble in water when present in the atmosphere in higher amount. During rain these react with water vapor of the atmosphere to form acids like sulphuric acid, sulphurous acid, nitric acid and nitrous acid. Building, bridges and other manmade structures can be soiled and damaged by acid rains and air pollution. The accumulation of dust can alter the appearance of a building. Sulphur dioxide in smog chemically transforms marble into gypsum, causing it to crack and flake off.

Ozone depletion

Ozone is present at all altitudes in the atmosphere mainly in the stratosphere (extending from 12 kms to 35 kms). This layer protects the earth's atmosphere from ultraviolet radiations coming from solar system. In nature, there is equilibrium between ozone and oxygen. Ozone depletion (decrease in concentration of ozone) involves the conversion of atmospheric nitrogen into NO_2 due to solar activity or anthropogenic processes. NO_2 and CFCs are ozone eaters. Ozone depletion causes passing of harmful cosmic rays and ultraviolet rays and other harmful rays to the troposphere and these rays affect human life. This will results in increase of skin cancer and increasing temperature.

ROLE OF BIOTECHNOLOGY IN AIR POLLUTION CONTROL

The surroundings around us are termed as 'environment'. Our environment includes the abiotic component (the non living) and biotic component (the living). The abiotic environment includes air, water and soil; and the biotic environment consists of all living organisms such as plants, animals and microorganisms. Environmental pollution broadly refers to the presence of undesirable substances in the environment which are harmful to man and other organisms. There has been a significant increase in the levels of harmful environmental pollution mostly due to direct or indirect human activities in recent past. The major sources of environmental pollution are industries, agricultural and other anthropogenic and biogenic sources etc. The pollutants are chemical, biological and physical in nature.

Controlling the environmental pollution and the conservation of environment and biodiversity and controlling environmental pollution are the major focus areas of all the countries around the world. In this context, the importance and impact of biotechnological approaches and the implications of biotechnology has to be thoroughly evaluated. There have been serious concerns regarding the use of biotechnological products and the impact assessment of these products due to their interaction with the environmental factors. A lobby of the environmentalists has expressed alarm on the release of genetically engineered organisms in the atmosphere and have stressed on thorough investigation and proper risk assessment of these organisms before releasing them in to the environment. The effect of the effluents from biotechnological companies is also a cause of concern for everyone. The need of the hour is to have a proper debate on the safety of the use of the biotechnological products. The efforts are not only on to use biotechnology to protect the environment from pollution but also to use it to conserve the natural resources. As we all know that microorganisms are known natural scavengers so the microbial preparations (both natural as well as genetically engineered) can be used to clean up the environmental hazards.

Biotechnology is being used to provide alternative cleaner technologies which help to further reduce the hazardous environmental implications of the traditional technologies. When applied to air filtration and purification, Biofilters use microorganisms to remove air pollution. The air flows through a packed bed and the pollutant transfers into a thin biofilm on the surface of the packing material. Microorganisms, including bacteria and fungi are immobilized in the biofilm and degrade the pollutant. Trickling filters and bioscrubbers rely on a biofilm and the bacterial action in their recirculating waters.

The technology finds greatest application in treating malodorous compounds and water-soluble volatile organic compounds (VOCs). Industries employing the technology include food and animal products, off-gas from wastewater treatment facilities, pharmaceuticals, wood products manufacturing, paint and coatings application and manufacturing and resin manufacturing and application etc. Compounds treated are typically mixed VOCs and various sulfur compounds, including hydrogen sulfide. Very large airflows may be treated and although a large area (footprint) has typically been required -- a large Biofilters (>200,000 acfm) may occupy as much or more land than a football field -- this has been one of the principal drawbacks of the technology. Engineered biofilters, designed and built since the early 1990s, have provided significant footprint reductions over the conventional flat-bed, organic media type.

Using Bioreactors to Control Air Pollution

Biofilters

In air pollution, bioreaction is simply the use of microbes to consume pollutants from a contaminated air stream. Almost any substance, with the help of microbes, will decompose (decay) given the proper environment. This is especially true for organic compounds. But certain microbes also can consume inorganic compounds such as hydrogen sulfide and nitrogen oxides. Bioreaction is a "green" process, whereas the traditional approaches are not. Combusting any fuel will generate oxides of nitrogen (NO_x), particulate matter, sulfur dioxide (SO_2) and carbon monoxide (CO). Bioreactors usually do not generate these pollutants or any hazardous pollutants. Products of a bioreaction consuming hydrocarbons are water and carbon dioxide (CO_2). Microbes have a simple life cycle; they are born, eat, grow, reproduce and die. Their diet is based primarily on carbon-based compounds, water, oxygen (for aerobic reactions) and macronutrients. Bioreactors use microbes to remove pollutants from emissions by consuming the pollutants.

Bioreactors have been used for hundreds of years to treat sewage and other odoriferous, water-borne waste. The initial process used a device called a "biofilter". A biofilter is usually a rectangular box that contains an enclosed plenum on the bottom, a support rack above the plenum, and several feet of media (bed) on top of the support rack. A large number of materials are used for bed media such as peat, composted yard waste, bark, coarse soil, gravel or plastic shapes. The support rack is perforated to allow air from the plenum to move into the bed media to contact microbes that live in the bed. The perforations also permit excess, condensed moisture to drain out of the bed to the plenum. A fan is used to collect contaminated air from a building or process. If the air is too hot, too cold, too dry, or too dirty (with suspended solids), it may be necessary to pre-treat the contaminated air stream to obtain optimum conditions before introducing it into a bioreactor **(Fig 3.2 (i) and Fig 3.2 (ii))**.

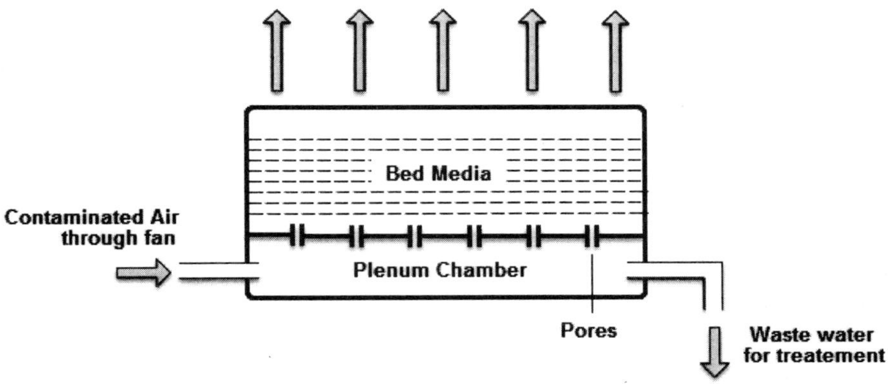

Fig. 3.2(i): Biofilter

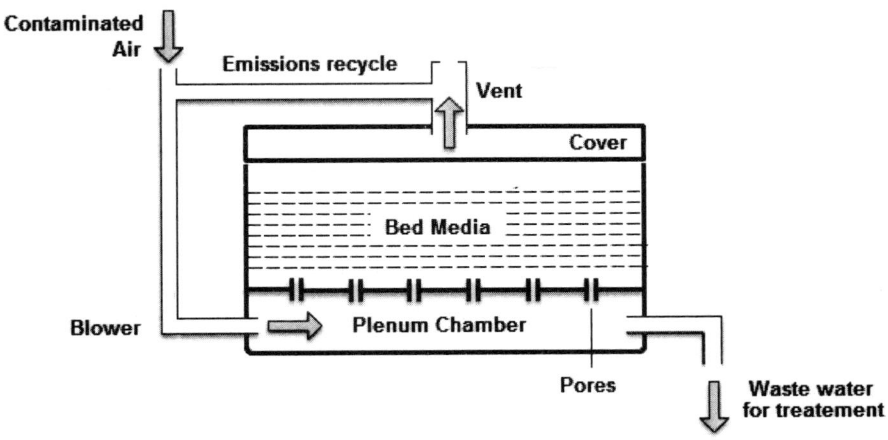

Fig. 3.2(ii): Biofilter with recycle process

Contaminated air is duct to a plenum. As the emissions flow through the bed media, the pollutants are absorbed by moisture on the bed media and come into contact with microbes. Microbes reduce pollutant concentrations by consuming and metabolizing pollutants. During the digestion process, enzymes in the microbes convert compounds into energy, CO_2 and water. Material that is indigestible is left over and becomes residue.

Factors affecting the Biofilters

Because bioreactors use living cultures, they are affected by many variables in their environment. Below are variables and limitations that affect the performance of all bioreactors, regardless of process type.

Temperature

The most important variable affecting bioreactor operations is temperature. A blast of hot air can totally kill a biomass faster than any other accident. Most microbes can survive and flourish in a temperature range of 30 to 41 degrees. It is important to monitor bed temperature at least daily, but every eight hours would be safer. A high temperature alarm on the emissions inlet is also a good safety precaution. When emissions from a process are too hot, operators often pass hot emissions through a humidifier that cools gases down by evaporative cooling. Besides the cooling effect, this process also increases the moisture content (humidifies emission stream), a desirable side effect. Although a blast of really hot air is the most lethal variable for microbes, cold air also stops, but does not kill, microbes. Cold air can reduce microbe activity to the point that they stop consuming pollutants and go into a state of suspended animation.

Moisture

The second most critical variable is bed moisture. Microbes need moisture to survive and moisture creates the bio-film that removes (absorbs) pollutants from an air stream so that they can be assimilated by microbes. Low moisture problems can be corrected by passing emissions through a humidifier. Having emissions close to saturation (100 % relative humidity) will solve most dry bed problems. Biofilters are usually operated damp with no running or standing water. Low moisture, for short periods, will not kill the microbes, but low moisture will greatly reduce efficiency. Efficiency will be below optimum while microbes recover (re-acclimate) after a period of dry bed conditions It is important to remember that a by-product of a bioreaction is water. If emissions are saturated entering the process, there will be water condensing in the bed media. Always provide space in the plenum for water to collect and a method to remove it from the plenum.

Nutrients

In addition to a comfortable temperature and a moist environment, microbes need a diet of balanced nutrients to survive and propagate. Pollutants provide the main source of food and energy, but microbes also require macronutrients to sustain life. Decay of an organic bed media can provide most macronutrients. However, if a bed is deficient in certain nutrients, microbes will cease to grow and could begin to die. Nitrogen is an essential nutrient for microbial growth. Microbes use nitrogen to build cell walls (these walls contain approximately 15 percent nitrogen) and nitrogen is a major constituent of proteins and nucleic acids. Other essential macronutrients include phosphorus, potassium, sulfur, magnesium, calcium, sodium and iron. Nitrogen, phosphorus, potassium (the NPK code on fertilizer labels) may be added by incorporating agricultural fertilizer into bed media. Lesser soluble macronutrients such as magnesium, calcium, sodium and iron, may be purchased in small quantities at feed and seed stores. The nutrient content of a bed should be checked periodically by submitting samples to a soils lab for analysis.

pH

Most bioreactors perform best when the bed pH is near 7, or neutral. Some pollutants form acids when they decompose. Examples of these compounds are: hydrogen sulfide, organic sulfur compounds, and halogens (chlorine, fluoride, bromine and iodine). Production of acids over time will lower pH and will eventually destroy microbes. If a process emits pollutants that produce acids, a plan must be developed to neutralize these acids. There are several techniques available to neutralize beds. Some may be incorporated into specification for the bed material. One of the simplest techniques is to mix oyster shells with bed media. The shells will eventually dissolve and have to be replaced. How long the shells last depends on how much acid is produced. Another simple technique is to install a garden soaking hoses in the packing media during construction. Periodically, a dilute solution of soda ash (sodium carbonate, Na_2CO_3) may be introduced into a bed when pH begins to decline. Another technique is to spray dilute soda ash solution over the top of the bed. However, this will probably be less effective than wetting the core of a bed with soaker hoses.

Type of Microbes

The strains that flourish on pollutants in an emission stream will eventually dominate the bed environment. The natural method will take a little longer to acclimate to optimum efficiency, but, because of the diversity of the strains of microbes, will be more adaptable in the long run. Specific microbes that are developed in the lab are more susceptible to changes in the environment than naturally generated microbes.

Biofilter Advantages

(i) Installation costs are low- Most biofilters are constructed from common materials locally available such as lumber, fiberglass, and plastic pipe. They can be assembled using carpenters, plumbers, and earthmovers.

(ii) Depending on the amount of pretreatment the emissions require operating costs are usually low. These costs consist of electricity to operate the primary blower and the humidification pump, part-time labor to check on the process, and small quantities of macronutrients.

(iii) Over time, some modifications have been developed to overcome some of the specific deficiencies in the traditional biofilter design. To increase contact time with microbes, some facilities recycle a portion of the exhaust back through the bioreactor. This is done by adding a cover and vent to the biofilter. A slipstream is taken from the vent and is recycled back to the intake of the primary blower. Biofilters modules may be added horizontally, in series **(Fig 3.3)**.

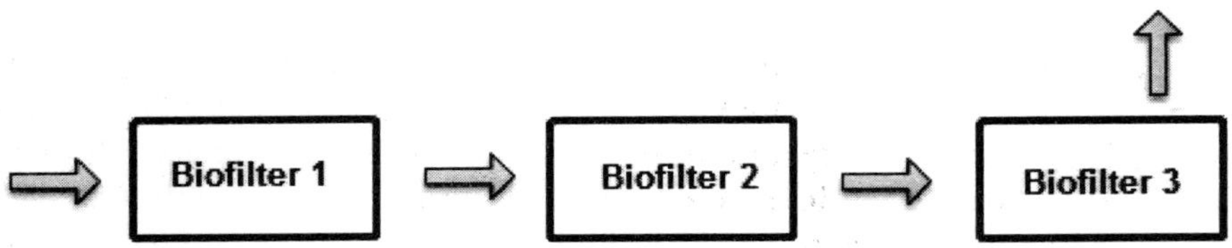

Fig. 3.3: Biofilters added in series

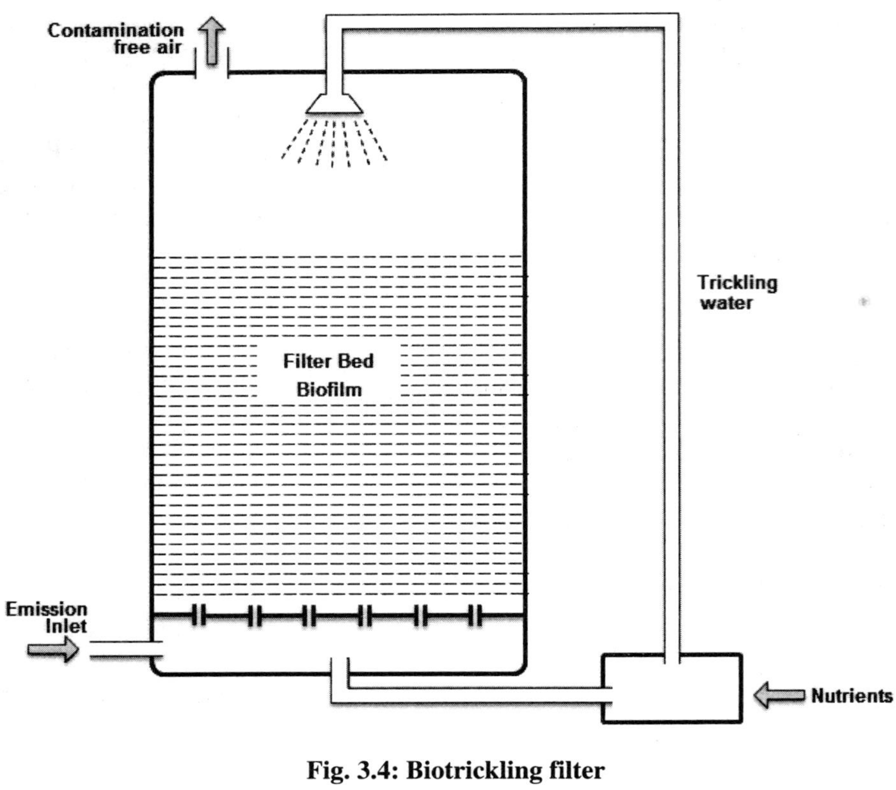

Fig. 3.4: Biotrickling filter

Biotrickling Filter

A trickling filter is a wastewater treatment process that is usually a round, vertical tank that contains a support rack and is filled with aggregate, ceramic or plastic media to a height of 3 to 15 feet. In the middle of the tank is a vertical pipe that has a rotary connection on the top end. A spray arm is attached to the rotary connection and this has spray nozzles installed along its length. The spray nozzles are angled slightly off-center to provide force necessary to rotate the spraying arm around the top of the trickling filter. A recirculating pump is used to pump liquid from the reservoir in the bottom to the spray nozzles. Liquid level in the sump is maintained with an automatic effluent make-up system. A biofilm forms on the packing surface. This is a biologically active mass that removes the pollutants from the effluent and the microbes decompose them **(Fig 3.4)**.

The Biotrickling filter is very similar to the trickling filter. However, the pollutants are contained in an air phase (emissions), and the pollutants must be dissolved into the liquid phase to be available to the microbes. As the air phase passes through the packing, the pollutants are absorbed from the air into the liquid phase to achieve maximum contact with the biomass. This is the difference from the trickling filter because pollutants that enter the system are already in the liquid phase (effluent) in the trickling filter. Water is added to the reservoir to make-up for water that has evaporated. Accumulated bio-sludge is periodically removed from the reservoir and disposed. Emissions may be routed through the Biotrickling filter co-current or counter-current to the effluent flow. Because of the continuous flow of a liquid phase, it is an easy matter to automatically neutralize acid build-up.

Bioscrubbers

Just as the Biotrickling filter is an enhancement of the biofilter, the bioscrubber is an enhancement to the biotrickling filter. The bioscrubber attempts to solve two problems with the biotrickling filter:

(i) Improve the absorption of pollutants into the liquid.

(ii) Lengthen the time.

The microbes have to consume the pollutants. These are accomplished in two ways: the tower packing is flooded with a liquid phase and the discharge effluent from the bioscrubber is collected in a storage tank before being recycled back to the bioscrubber. Bioscrubbers are an odor treatment technology that utilizes biological processes, as opposed to chemical processes as their treatment mechanisms. They are similar to biofilters in this regard. Bioscrubbers use artificial media and closed vessel construction, where biofilters often use natural media and open bed construction.

The process involves intermittent spraying or recirculating biologically active, nutrient rich scrubbing solutions over an artificial media while odorous air is forced upward through the media bed. The process is similar to that used in wet scrubbers, except it involves biological treatment instead of chemical treatment. The media provide sites for biological colonization and promote mass transfer from the air to the water film on the biomass where the biological oxidation occurs **(Fig 3.5)**.

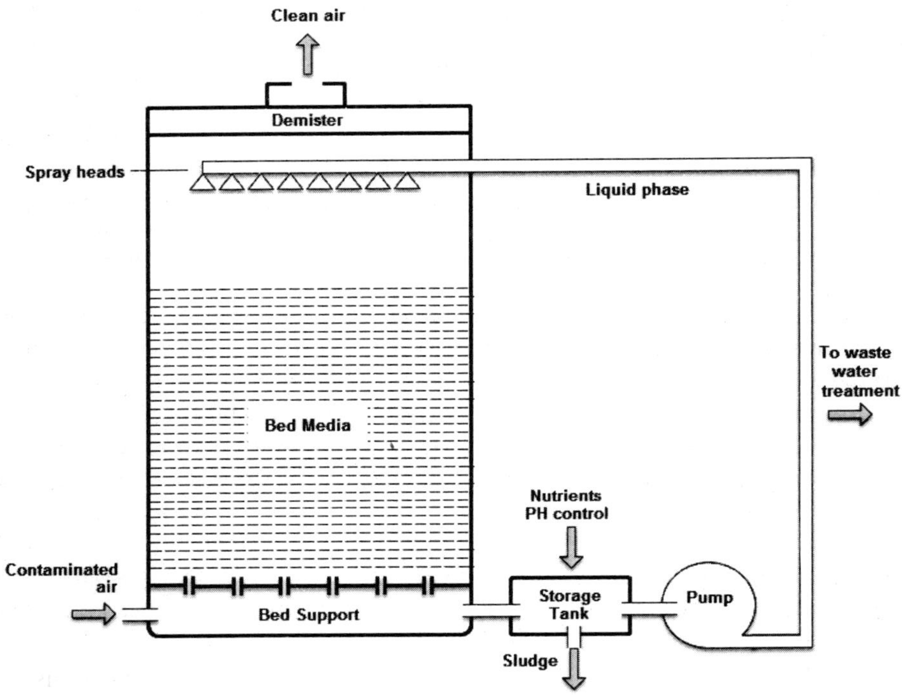

Fig. 3.4: Bioscrubber

Flooding the bed increases the ability of the liquid phase to absorb pollutants because as the gas phase (emissions) impacts the bed media it forms tiny bubbles that greatly increases the surface-area of the interface between the gas and liquid phases. Increasing the interface-area improves the liquid phase's ability to absorb pollutants.

The storage tank acts as reservoir for the liquid phase and permits additional reaction time for the microbes to consume pollutants. Reaction times can be increased to an hour or more, depending on the recirculation rate of the liquid phase and the size of the sump. This increases the time available for microbes to attach and destroy pollutants.

Major Design Considerations

Irrigation

The media must be kept moist, but if irrigation is too frequent, the biomass may be deprived of oxygen. If irrigation is too infrequent, the media can dry out and reduce effectiveness. A programmable timer may be used to properly time irrigation cycles.

Media

Media should possess high surface area to volume ratios, good adsorption characteristics, low pressure drop, and good sloughing characteristics. For organic compound treatment, media impregnated with adsorptive materials such as activated carbon may be used.

Nutrients

Nutrients must be kept fresh to maintain healthy biomass. Appropriate nutrients essential to the various bacteria in the bioscrubber should be identified. Wastewater treatment plant effluent is generally an excellent source for these nutrients. Nutrient mixes may also be purchased from manufacturers and mixed into the bioscrubber solution.

Air distribution

Uniform air distribution through the media bed is important to efficient operation. Perforated FRP distribution plates have been used effectively to support the media bed and distribute the air flow uniformly.

Bioscrubber Advantages

a) It is not necessary to humidify emissions prior to treating them. This could save the cost of installing a humidification process.

b) The bioscrubber has a smaller footprint than other bioreactors. This is an important consideration in congested facilities with limited available real estate.

c) Because pH control and nutrient feed can be automated, it requires less attention than other bioreactors.

d) Process is ideal for emissions that produce acids upon treatment.

e) Bioscrubber can treat emissions containing particulate matter.

Bioscrubber Disadvantages

a) Considerably more expensive to install than other bioreactors. It has a chemical scrubber at the heart of the process and resembles chemical-processing equipment more so than other bioreactors.

b) Over feeding can cause excessive biomass growth, which can plug the bioscrubber.

c) Operating cost can be higher than other bioreactor processes.

d) Needs expensive and complex feeding and neutralizing systems.

e) To control biomass growth, toxic and dangerous compounds must be inventoried and handled.

Biofiltration will play a major role in the treatment of organic and inorganic emissions from a variety of industrial and waste water treatment processes. Biofiltration, when compared to other available technologies, has significant technical and cost advantages. Compost biofilters are better suited for treatment of odors and low concentration (< 25 ppmv) contaminants. Biotrickling filters have significant advantages over compost biofilters and are capable of handling significantly higher contaminant concentrations (20 ppmv – 5,000 ppmv). The major issues in biotrickling filters are the design of the support media and handling of biomass growth. Support media design has a significant impact on biotrickling filter performance. The market for biofilters will increase in the next millennium, as new applications arise in the future.

4

Waste Water Management

Clean and plentiful water provides the foundation for prosperous communities. We rely on clean water to survive, yet right now we are heading towards a water crisis. Changing climate patterns are threatening lakes and rivers, and key sources that we tap for drinking water are being overdrawn or tainted with pollution. Safe drinking water is essential to humans and other life forms even though it provides no calories or organic nutrients. Access to safe drinking water has improved over the last decades in almost every part of the world, but approximately one billion people still lack access to safe water and over 2.5 billion lack access to adequate sanitation. Water plays an important role in the world economy, as it functions as a solvent for a wide variety of chemical substances and facilitates industrial cooling and transportation. Approximately 70% of the fresh water used by humans goes to agriculture.

This natural resource is becoming scarcer in certain places, and its availability is a major social and economic concern. Currently, about a billion people around the world routinely drink unhealthy water. Poor water quality and bad sanitation are deadly; some five million deaths a year are caused by polluted drinking water. The World Health Organization estimates that safe water could prevent 1.4 million child deaths from diarrhea each year.

WATER POLLUTION

Dirty water is the world's biggest health risk, and continues to threaten both quality of life and public health in world. When toxic substances enter lakes, streams, rivers, oceans, and other water bodies, they get dissolved or lie suspended in water or get deposited on the bed. This results in the pollution of water whereby the quality of the water deteriorates, affecting aquatic ecosystems. Pollutants can also seep down and affect the groundwater deposits. Water pollution has many sources. The most polluting of them are the city sewage and industrial waste discharged into the rivers. Due to this, pollutants enter groundwater, rivers, and other water bodies. Such water, which ultimately ends up in our households, is often highly contaminated and carries disease-causing microbes. Agricultural run-off, or the water from the fields that drains into rivers, is another major water pollutant as it contains fertilizers and pesticides.

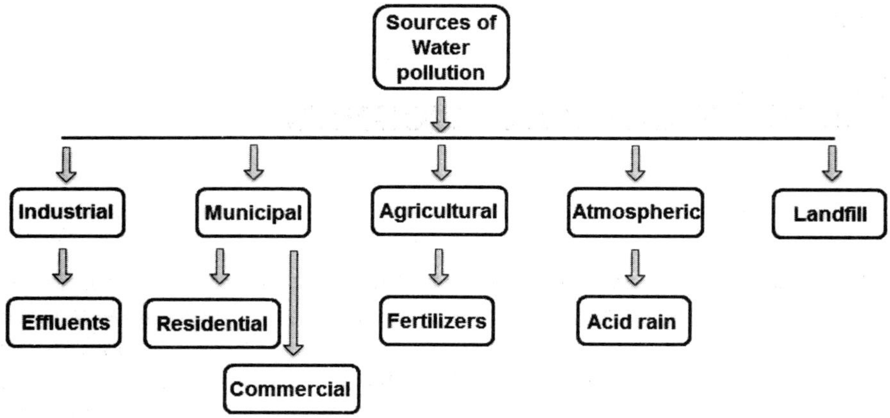

Fig. 4.1: Sources of water pollution

MAJOR SOURCES OF WATER POLLUTION

Wastewater is not just sewage. All the water used in homes, restaurants, businesses, and industries that goes down the drains or into the sewage collection system is wastewater. This includes water from baths, showers, sinks, dishwashers, washing machines, and toilets. Domestic sewage amounts to a very small fraction of the sewage by weight. But it is large by volume and contains impurities such as organic materials and plant nutrients that tend to rot. The main organic materials are food and vegetable waste, plant nutrient come from chemical soaps, washing powders, etc. Domestic sewage is also very likely to contain disease-causing microbes. Thus, disposal of domestic waste water is a significant technical problem **(Fig4.1)**.

Domestic Sewage

Today, many people dump their garbage into streams, lakes, rivers, and seas, thus making water bodies the final resting place of cans, bottles, plastics, and other household products. The various substances that we use for keeping our houses clean add to water pollution as they contain harmful chemicals. In the past, people mostly used soaps made from animal and vegetable fat for all types of washing. But most of today's cleaning products are synthetic detergents and come from the petrochemical industry. Most detergents and washing powders contain phosphates, which are used to soften the water among other things. These and other chemicals contained in washing powders affect the health of all forms of life in the water.

Agricultural Pollutants

The use of land for agriculture and the practices followed in cultivation greatly affect the quality of groundwater. Intensive cultivation of crops causes chemicals from fertilizers (e.g. nitrate) and pesticides to seep into the groundwater, a process commonly known as leaching. Routine applications of fertilizers and pesticides for agriculture and indiscriminate disposal of industrial and domestic wastes are increasingly being recognized as significant sources of water pollution. The high nitrate content in groundwater is mainly from irrigation run-off from agricultural fields where chemical fertilizers have been used indiscriminately.

Industrial Effluents

Waste water from manufacturing or chemical processes in industries contributes to water pollution. Industrial waste water usually contains specific and readily identifiable chemical compounds. During the last fifty years, the number of industries in India has grown rapidly. But water pollution is concentrated within a few subsectors, mainly in the form of toxic wastes and organic pollutants. Out of this a large portion can be traced to the processing of industrial chemicals and to the food products industry. Most of these defaulting industries are sugar mills, distilleries, leather processing industries, and thermal power stations. Some of the main substances released in water from industries are as follows:

1) Sulphur – This is a non-metallic substance that is harmful for marine life.

2) Asbestos – This pollutant has cancer-causing properties. When inhaled, it can cause illnesses such as asbestosis and many types of cancer.

3) Lead and Mercury – These are metallic elements and can cause environmental and health problems for humans and animals.

4) Nitrates & Phosphates– These are found in fertilizers are often washed from the soils to nearby water bodies. They can cause eutrophication, which can be very problematic to marine environments.

5) Oils – Oils forms a thick layer on the water surface because they do not dissolve in water. This can stop marine plants receiving enough light for photosynthesis. It is also harmful for fish and marine birds.

Underground Storage and Tube Leakages

Many liquid products (petroleum products) are stored in metal and steel tubes underground. Other sewage systems run in underground tubes. Overtime, they rust and begin to leak. If that happens, they contaminate the soils and the liquids in them end up in many nearby water bodies.

Atmospheric Sources

Atmospheric deposition is the pollution of water bodies caused by air pollution. Each time the air is polluted with sulphur dioxide and nitrogen oxide, they mix with water particles in the air and form a toxic substance. This falls as acid rain to the ground, and gets washed into water bodies. The result is that, water bodies also get contaminated and this affects animals and water organisms.

COMPOSITION OF WASTE WATER

The composition of wastewater varies widely. It can be summarized as follows:

a) Water (> 95%) which is often added during flushing to carry waste down a drain.

b) Pathogens such as bacteria, viruses, prions and parasitic worms.

c) Non-pathogenic bacteria.

d) Organic particles such as feces, hairs, food, vomit, paper fibers, plant material, humus, etc.

e) Soluble organic material such as urea, fruit sugars, soluble proteins, drugs, pharmaceuticals, etc.

f) Inorganic particles such as sand, grit, metal particles, ceramics, etc.

g) Soluble inorganic material such as ammonia, road-salt, sea-salt, cyanide, hydrogen sulfide, thiocyanates, thiosulfates, etc.

h) Animals such as protozoa, insects, arthropods, small fish, etc.

i) Macro-solids such as sanitary napkins, nappies/diapers, needles, children's toys, dead animals or plants, etc.

j) Gases such as hydrogen sulfide, carbon dioxide, methane, etc.

k) Emulsions such as paints, adhesives, mayonnaise, hair colorants, emulsified oils, etc.

l) Toxins such as pesticides, poisons, herbicides, etc.

m) Pharmaceuticals and hormones.

WATER CYCLE AND POLLUTION

The earth's water is constantly on the move in a circle of events called the water cycle **(Fig 4.2)**. On a warm day, water from the surface of the ocean, a pond or even a puddle evaporates and rises into the atmosphere. As water vapor rises, it cools and condenses to form clouds. The clouds become heavy with water until they can hold no more. Then raindrops form and fall. On land, the water runs downhill, eventually finding its way to a stream or river, then back to a lake or to the ocean where the cycle begins again. As the water runs off of the land and back to the oceans, it carries with it many substances that it picks up along its way. It could run through a pool of gasoline spilled at the pump, loose dirt from the construction site or waste from industries, or fertilizers from the agricultural fields. All of these cause pollution in the aquatic environment.

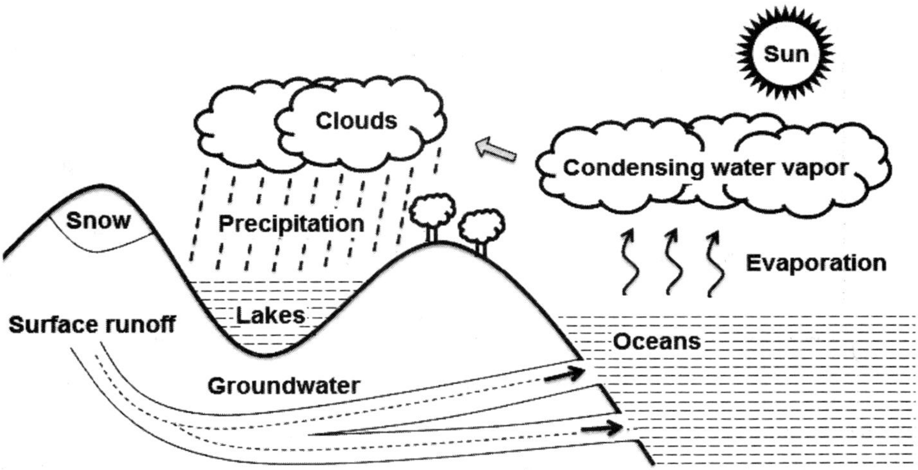

Fig. 4.2: Water cycle

Effects of Water Pollution

The effects of water pollution are not only devastating to people but also to animals, fish, and birds. Polluted water is unsuitable for drinking, recreation, agriculture, and industry. It diminishes the aesthetic quality of lakes and rivers. More seriously, contaminated water destroys aquatic life and reduces its reproductive ability. Eventually, it is a hazard to human health. Nobody can escape the effects of water pollution. It has serious impact on the water cycle **(Fig 4.3)**.

There are many different types of water pollution and all have a different adverse effect on the environment:

a) Heavy metals from industrial processes can accumulate in nearby lakes and rivers. These are toxic to marine life such as fish and shellfish, and can affect the rest of the food chain. This means that entire animal communities can be badly affected by this type of pollutant.

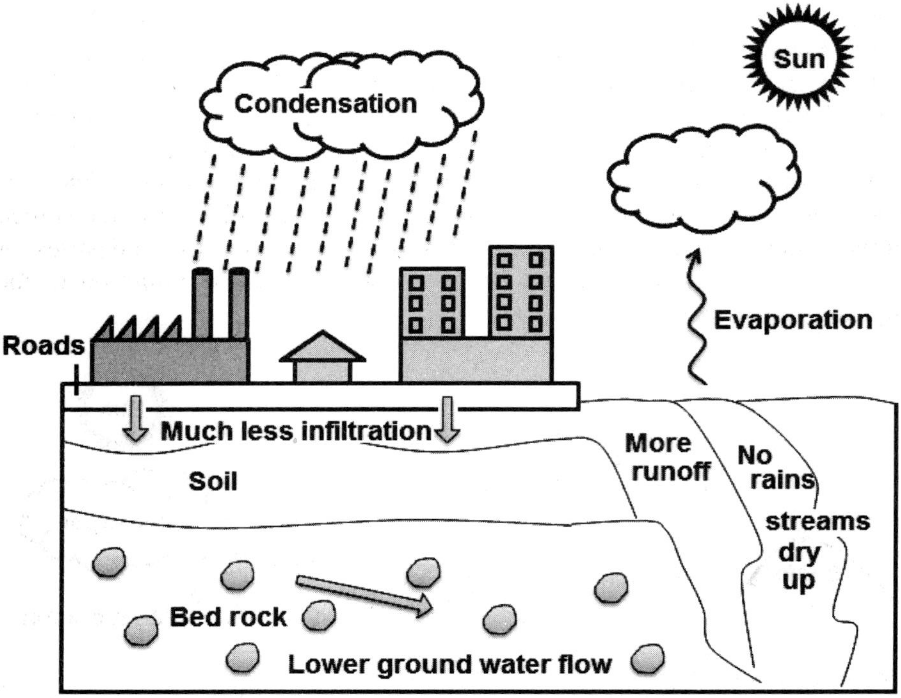

Fig. 4.3: Urban water cycle

b) Industrial waste often contains many toxic compounds that damage the health of aquatic animals and those who eat them. Some toxins affect the reproductive success of marine life and can therefore disrupt the community structure of an aquatic environment.

c) Microbial pollutants from sewage often result in infectious diseases that infect aquatic life and terrestrial life through drinking water. This often increases the number of mortalities seen within an environment.

d) Organic matter and nutrients causes an increase in aerobic algae and depletes oxygen from the water column. This is called eutrophication and causes the suffocation of fish and other aquatic organisms.

e) Sulfate particles from acid rain change the pH of water making it more acidic, this damages the health of marine life in the rivers and lakes it contaminates, and often increases the number of mortalities within an environment.

f) Suspended particles can often reduce the amount of sunlight penetrating the water, disrupting the growth of photosynthetic plants and micro-organisms. This has subsequent effects on the rest of the aquatic community that depend on these organisms to survive.

g) Ecosystems are destroyed by the rising temperature in the water, as coral reefs are affected by the bleaching effect due to warmer temperatures. Additionally, the warm water forces indigenous water species to seek cooler water in other areas, causing an ecological damaging shift of the affected area.

Human Health Effects

Water pollution can pose health dangers to humans who come into contact with it, either directly or indirectly. A large number of chemicals that either exist naturally in the land or are added due to human activity dissolve in the water, thereby contaminating it and leading to various diseases.

a) The organophosphates and the carbonates present in pesticides affect and damage the nervous system and can cause cancer. They contain chlorides that cause reproductive and endocrinal damage.

b) Lead is hazardous to health as it accumulates in the body and affects the central nervous system. Children and pregnant women are most at risk.

c) Excess fluorides can cause yellowing of the teeth and damage to the spinal cord and other crippling diseases.

d) Drinking water that gets contaminated with nitrates can prove fatal especially to infants that drink formula milk as it restricts the amount of oxygen that reaches the brain causing the 'blue baby' syndrome.

e) Benzene and other petrochemicals can cause cancer even at low exposure levels.

f) Arsenic poisoning through water can cause liver and nervous system damage, vascular diseases and also skin cancer.

g) Exposure to polluted water can cause diarrhea, skin irritation, respiratory problems, and other diseases, depending on the pollutant that is in the water body. Stagnant water and other untreated water provide a habitat for the mosquito and a host of other parasites and insects that cause a large number of diseases especially in the tropical regions. Among these, malaria is undoubtedly the most widely distributed and causes most damage to human health.

Eutrophication

When fresh water is artificially supplemented with nutrients, it results in an abnormal increase in the growth of water plants. This is known as eutrophication. The discharge of waste from industries, agriculture, and urban communities into water bodies generally stretches the biological capacities of aquatic systems **(Fig 4.4)**. Chemical run-off from fields also adds nutrients to water. Excess nutrients cause the water body to become choked with organic substances and organisms. When organic matter exceeds the capacity of the micro-organisms in water that break down and recycle the organic matter, it encourages rapid growth, or blooms, of algae. When they die, the remains of the algae add to the organic wastes already in the water; eventually, the water becomes deficient in oxygen. Anaerobic organisms (those that do not require oxygen to live) then attack the organic wastes, releasing gases such as methane and hydrogen sulphide, which are harmful to the oxygen-requiring (aerobic) forms of life. The result is a foul-smelling, waste-filled body of water.

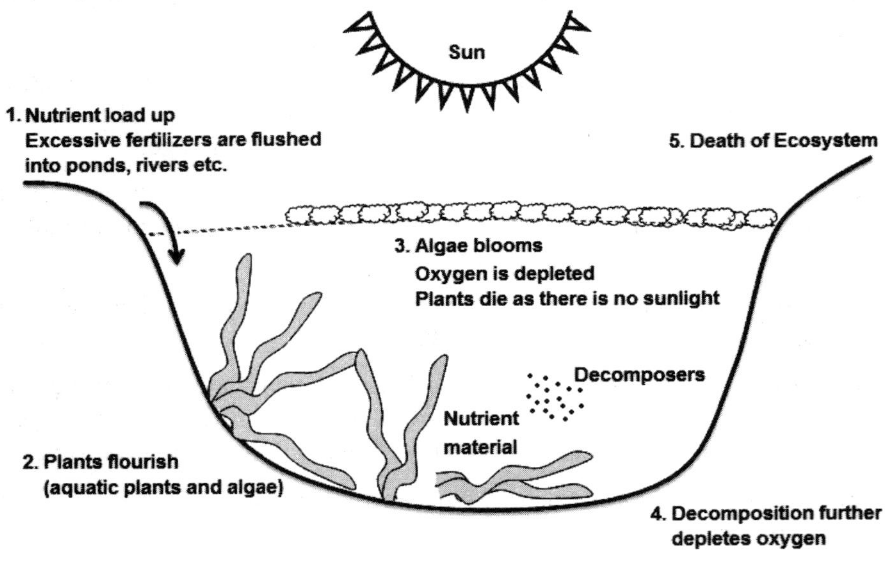

Fig. 4.4: Eutrophication

WASTE WATER QUALITY INDICATORS

There are various tests which indicate the pollution level in water. These tests measure physical, chemical, and biological characteristics of the wastewater. They are as follows:

a) Temperature- there is a specific temperature required by organisms living in water.

b) Solids- total dissolved or suspended solids give the indication of the level of pollution in water.

c) pH –it depicts the level of acidity in water.

d) Dissolved oxygen/ BOD- Dissolved oxygen concentrations may be measured directly in wastewater, but the amount of oxygen potentially required by other chemicals in the wastewater is termed an oxygen demand. Dissolved or suspended oxidizable organic material in wastewater will be used as a food source. Finely divided material is readily available to microorganisms whose populations will increase to digest the amount of food available. Digestion of this food requires oxygen, so the oxygen content of the water will ultimately be decreased by the amount required to digest the dissolved or suspended food. Oxygen concentrations may fall below the minimum required by aquatic animals if the rate of oxygen utilization exceeds replacement by atmospheric oxygen.

Biochemical Oxygen Demand

The amount of organic material that can rot in the water is measured by the biochemical oxygen demand. BOD is the amount of oxygen required by micro-organisms to decompose the organic substances in water. Therefore, the more organic material there is in the water, the higher the BOD. It is among the most important parameters for the design and operation of waste water treatment plants. BOD levels of industrial effluent may be many times that of domestic sewage. Dissolved oxygen is an important factor that determines the quality of water in lakes and rivers. The higher the concentration of dissolved oxygen, the better the water quality. When sewage enters a lake or stream, micro-organisms begin to decompose the organic materials. Oxygen is consumed as micro-organisms use it in their metabolism. This can quickly deplete the available oxygen in the water. When the dissolved oxygen levels drop too low, many aquatic species perish. In fact, if the oxygen level drops to zero, the water will become septic. When organic compounds decompose without oxygen, it gives rise to the undesirable odors usually associated with septic or putrid conditions.

The chemical oxygen demand (COD) test is commonly used to indirectly measure the amount of organic compounds in water. Most applications of COD determine the amount of organic pollutants found in surface water (e.g. lakes and rivers) or wastewater, making COD a useful measure of water quality. It is expressed in milligrams per liter (mg/L) also referred to as ppm (parts per million), which indicates the mass of oxygen consumed per liter of solution.

e) Nitrogen - Nitrogen is an important nutrient for plant and animal growth. Atmospheric nitrogen is less biologically available than dissolved nitrogen in the form of ammonia and nitrates. Availability of dissolved nitrogen may contribute to algal blooms. Ammonia and organic forms of nitrogen are often measured as Total Kjeldahl

Nitrogen, and analysis for inorganic forms of nitrogen may be performed for more accurate estimates of total nitrogen content.

f) Chlorine - Chlorine has been widely used for bleaching, as a disinfectant and for biofouling prevention in water cooling systems. Remaining concentrations of oxidizing hypochlorous acid and hypochlorite ions may be measured as chlorine residual to estimate effectiveness of disinfection or to demonstrate safety for discharge to aquatic ecosystems.

g) Biological characteristics - Water may be tested by a bioassay comparing survival of an aquatic test species in the wastewater in comparison to water from some other source. Water may also be evaluated to determine the approximate biological population of the wastewater. Pathogenic micro-organisms using water as a means of moving from one host to another may be present in sewage. Coli form index measures the population of an organism commonly found in the intestines of warm-blooded animals as an indicator of the possible presence of other intestinal pathogens.

WASTE WATER MANAGEMENT

There are two general treatment objectives with respect to wastewater:

a) Reducing or minimizing the public health hazards of a wastewater. These are general treatment measures aimed at preventing pathogens and other potentially harmful components from finding their way back to the consumer.

b) Eliminating, reducing or minimizing the deteriorative impact of a wastewater on the receiving water quality and its environment.

A sharp distinction must be made between the term "wastewater disposal" and "wastewater treatment". All wastewater has to be disposed of. Some wastewater is subjected to various types of treatment before disposal, but some wastewater receives no treatment before disposal.

Wastewater Disposal

There are three methods by which final disposal of wastewater can be accomplished. The general problem areas that are of concern in final disposal are pathogenic microorganisms (viruses, etc.), heavy metals and the presence of biologically resistant organic compounds, such as pesticides or insecticides which can find their way into water supplies. More recently, there has been interest in the use of land for both surface and subsurface disposal after wastewater treatment.

Surface Disposal

Generally this is disposal by irrigation. This involves spreading the wastewater over the surface of the ground, generally by irrigation ditches. There is some evaporation, but most of the wastewater soaks into the ground and supplies moisture with small amounts of fertilizing ingredients for plant life. This method is largely restricted to small volumes of wastewater from a relatively small population where land area is available and where nuisance problems will not be created. It has its best use in arid or semi-arid areas where the moisture added to the soil is of special value. If crops are cultivated on the disposal area, the growth of vegetation often must be excluded from wastewater. Because untreated wastewater will also contain pathogenic organisms, the production of foods for human consumption which may be eaten without cooking is not desirable.

Subsurface Disposal

By this method wastewater is introduced into the ground below its surface through pits or tile fields. It is commonly used for disposal of settled wastewater from residences or institutions where there is only a limited volume of wastewater.

Disposal by Dilution

Disposal by dilution is the simple method of discharging wastewater into surface water such as a river, lake, ocean, estuaries or wetlands. This results in the pollution of the receiving water. The degree of pollution depends on the dilution, volume and composition of the wastewater as compared to the volume and quality of the water with which it is mixed. When the volume and organic content of the wastewater is small, compared with the volume of the receiving water, the dissolved oxygen present in the receiving water is adequate to provide for aerobic decomposition of the organic solids in the wastewater so that nuisance conditions do not develop. However, in spite of the continued aerobic status of the receiving water, microbial pollution remains a health menace and floating solids in the wastewater, if not previously removed, is visible evidence of the pollution.

Less obvious problems associated with this type of disposal are the effects of toxic or potentially toxic compounds found in domestic and industrial wastewater. These may involve immediate toxic effects such as heavy metals in fish and the "concentration" of certain biologically resistant compounds in the food chain. An example would be the accumulation of certain pesticides by microorganisms that are consumed by higher organisms to include fish, birds, and even man. Another subtle environmental effect now of some concern due to disposal of untreated wastewater by dilution is the enrichment of receiving waters by the introduction of plant nutrients such as nitrogen and phosphorous. The presence of excessive amounts of these nutrients can stimulate plant and algae growth in the receiving waters. This is of special concern in inland, enclosed waters such as lakes and ponds, where the body of water can be harmed by gradual long-range changes that take place over the years.

NEED FOR WASTEWATER TREATMENT

In addition to water that we want to recycle, wastewater contains pathogens (disease organisms), nutrients such as nitrogen and phosphorus, solids, chemicals from cleaners and disinfectants and even hazardous substances. Given all of the components of wastewater, it seems fairly obvious that we need to treat wastewater not only to recycle the water and nutrients but also to protect human and environmental health. Satisfactory disposal of wastewater, whether by surface, subsurface methods or dilution, is dependent on its treatment prior to disposal. Adequate treatment is necessary to prevent contamination of receiving waters to a degree which might interfere with their best or intended use, whether it be for water supply, recreation, or any other required purpose.

Usually wastewater treatment will involve collecting the wastewater in a central, segregated location (the Wastewater Treatment Plant) and subjecting the wastewater to various treatment processes. Most often, since large volumes of wastewater are involved, treatment processes are carried out on continuously flowing wastewaters (continuous flow or "open" systems) rather than as "batch" or a series of periodic treatment processes in which treatment is carried out on parcels or "batches" of wastewaters. While most wastewater treatment processes are continuous flow, certain operations, such as vacuum filtration, involving as it does storage of sludge, the addition of chemicals, filtration and removal or disposal of the treated sludge, are routinely handled as periodic batch operations.

Wastewater treatment, however, can also be organized or categorized by the nature of the treatment process operation being used; for example, physical, chemical or biological. A complete treatment system may consist of the application of a number of physical, chemical and biological processes to the wastewater.

Physical Methods of Waste Water Treatment

Physical methods of wastewater treatment include sedimentation, flotation, and adsorption, as well as barriers such as bar racks, screens, deep bed filters, and membranes. Physical methods include processes where no gross chemical or biological changes are carried out and strictly physical phenomena are used to improve or treat the wastewater.

Screening

The influent sewage water passes through a bar screen to remove all large objects like cans, rags, sticks, plastic packets etc. carried in the sewage stream. This is most commonly done with an automated mechanically raked bar screen in modern plants serving large populations, whilst in smaller or less modern plants; a manually cleaned screen may be used. The solids are collected and later disposed in a landfill, or incinerated.

Sedimentation

In the process of sedimentation, physical phenomena relating to the settling of solids by gravity are allowed to operate. Usually this consists of simply holding a wastewater for a short period of time in a tank under quiescent conditions, allowing the heavier solids to settle, and removing the "clarified" effluent. Sedimentation for solids separation is a very common process operation and is routinely employed at the beginning and end of wastewater treatment operations.

Aeration and Filtration

Another physical treatment process consists of aeration -- that is, physically adding air, usually to provide oxygen to the wastewater. Still other physical phenomena used in treatment consist of filtration. Here wastewater is passed through a filter medium to separate solids. An example would be the use of sand filters to further remove entrained solids from a treated wastewater.

Equalization

In certain industrial wastewater treatment processes strong or undesirable wastes are sometimes produced over short periods of time. Since such "slugs" or periodic inputs of such wastes would damage a biological treatment process, these wastes are sometimes held, mixed with other wastewaters, and gradually released, thus eliminating "shocks" to the treatment plant. This is called equalization.

Chemical Methods of Waste Water Treatment

Chemicals are used during wastewater treatment in an array of processes to expedite disinfection. These chemical processes, which induce chemical reactions, are called chemical unit processes, and are used alongside biological and physical cleaning processes to achieve various water standards. There are several distinct chemical unit processes, including chemical coagulation, chemical precipitation, chemical oxidation and advanced oxidation, ion exchange, and chemical neutralization and stabilization, which can be applied to wastewater during cleaning.

Chemical Precipitation

Chemical precipitation is the most common method for removing dissolved metals from wastewater solution containing toxic metals. To convert the dissolved metals into solid particle form, a precipitation reagent is added to the mixture. A chemical reaction, triggered by the reagent, causes the dissolved metals to form solid particles. Filtration can then be used to remove the particles from the mixture. How well the process works is dependent upon the kind of metal present, the concentration of the metal, and the kind of reagent used. In hydroxide precipitation, a commonly used chemical precipitation process, calcium or sodium hydroxide is used as the reagent to create solid metal hydroxides. However, it can be difficult to create hydroxides from dissolved metal particles in wastewater because many wastewater solutions contain mixed metals.

Chemical Coagulation

This chemical process involves destabilizing wastewater particles so that they aggregate during chemical flocculation. Fine solid particles dispersed in wastewater carry negative electric surface charges (in their normal stable state), which prevent them from forming larger groups and settling. Chemical coagulation destabilizes these particles by introducing positively charged coagulants that then reduce the negative particles' charge. Once the charge is reduced, the particles freely form larger groups. Next, an anionic flocculent is introduced to the mixture. Because the flocculent reacts against the positively charged mixture, it either neutralizes the particle groups or creates bridges between them to bind the particles into larger groups. After larger particle groups are formed, sedimentation can be used to remove the particles from the mixture.

Chemical Oxidation and Advanced Oxidation

With the introduction of an oxidizing agent during chemical oxidation, electrons move from the oxidant to the pollutants in wastewater. The pollutants then undergo structural modification, becoming less destructive compounds. Alkaline chlorination uses chlorine as an oxidant against cyanide. However, alkaline chlorination as a chemical oxidation process can lead to the creation of toxic chlorinated compounds, and additional steps may be required. Advanced oxidation can help remove any organic compounds that are produced as a byproduct of chemical oxidation, through processes such as steam stripping, air stripping, or activated carbon adsorption.

Ion Exchange

When water is too hard, it is difficult to use to clean and often leaves a grey residue. (This is why clothing washed in hard water often retains a dingy tint.) An ion exchange process can be used to soften the water. Calcium and magnesium are common ions that lead to water hardness. To soften the water, positively charged sodium ions are introduced in the form of dissolved sodium chloride salt, or brine. Hard calcium and magnesium ions exchange places with sodium ions, and free sodium ions are simply released in the water. However, after softening a large amount of water, the softening solution may fill with excess calcium and magnesium ions, requiring the solution be recharged with sodium ions.

Chemical Stabilization

This process works in a similar fashion as chemical oxidation. Sludge is treated with a large amount of a given oxidant, such as chlorine. The introduction of the oxidant slows down the rate of biological growth within the sludge, and also helps deodorize the mixture. The water is then removed from the sludge. Hydrogen peroxide can also be used as an oxidant, and may be a more cost-effective choice.

Biological Treatment Methods

Biological treatment is an important and integral part of any wastewater treatment plant that treats wastewater from either municipality or industry having soluble organic impurities or a mix of the two types of wastewater sources. The obvious economic advantage, both in terms of capital investment and operating costs, of biological treatment over other treatment processes like chemical oxidation; thermal oxidation etc. has cemented its place in any integrated wastewater treatment plant.

Biological wastewater treatment, in its simplest form, is the conversion of biodegradable waste products from municipal or industrial sources by biological means. The practice of using a controlled biological population to degrade waste has been used for centuries, but while early wastewater treatment processes were quite simple, they have become more and more complex over time. Biological treatment methods use microorganisms, mostly bacteria, in the biochemical decomposition of wastewaters to stable end products. More microorganisms, or sludges, are formed and a portion of the waste is converted to carbon dioxide, water and other end products. Generally, biological treatment methods can be divided into aerobic and anaerobic methods **(Table 4.1)**, based on availability of dissolved oxygen.

Aerobic Methods of Waste Water Treatment

Biological wastewater treatment is an extremely cost effective and energy efficient system for the removal of BOD (Biological Oxygen Demand), since only, naturally occurring, micro-organisms are used. These micro-organisms feed on the complex materials present in the wastewater and convert them into simpler substances, preparing the water for further treatment. Aerobic wastewater treatment is a biological process that takes place in the presence of oxygen. Conventional aerobic treatment consists of an activated sludge process or oxidation lagoons. The size of these can be reduced, and tolerances against fluctuations in loads and toxics can be controlled in various ways, such as, by integrating moving bed bioreactors into the activated sludge treatment. As technology advances, systems are becoming increasingly convenient, more efficient, and smaller in size.

Table 4.1 Comparison of aerobic and anaerobic waste water treatment methods

PARAMETER	AEROBIC TREATMENT	ANAEROBIC TREATMENT
Process Principle	• Microbial reactions take place in the presence of molecular/ free oxygen • Reactions products are carbon dioxide, water and excess biomass	Microbial reactions take place in the absence of molecular/ free oxygen • Reactions products are carbon dioxide, methane and excess biomass
Applications	Wastewater with low to medium organic impurities (COD < 1000 ppm) and for wastewater that are difficult to biodegrade e.g. municipal sewage, refinery wastewater etc.	Wastewater with medium to high organic impurities (COD > 1000 ppm) and easily biodegradable wastewater e.g. food and beverage wastewater rich in starch/sugar/ alcohol.
Reaction Kinetic	Relatively fast	Relatively slow
Post Treatment	Typically direct discharge or filtration/disinfection	Invariably followed by aerobic treatment
Example Technologies	Activated Sludge e.g. Extended Aeration, Oxidation Ditch, MBR, Fixed Film Processes e.g. Trickling Filter/Biotower	Continuously stirred tank reactor/digester, Upflow Anaerobic sludge Blanket (UASB), Ultra High Rate Fluidized Bed reactors etc.

Biological Anaerobic Treatment

Anaerobic wastewater treatment is the biological treatment of wastewater without the use of air or elemental oxygen. In anaerobic treatment, organic pollutants are converted by anaerobic micro-organisms to a gaseous product that has the potential for reuse. Biological anaerobic treatment is a very low energy process that produces a fraction of the waste sludge of aerobic biological processes and is ideal for treating wastewater which is high in soluble BOD and/or COD.

Trickling Filters (TF)

The Trickling Filter is an aerobic treatment system that utilizes microorganisms attached to a medium to remove organic matter from wastewater. In most wastewater treatment systems, the trickling filter follows primary treatment and includes secondary settling tank or clarifier. The process is a fixed film biological treatment method designed to remove BOD and suspended solids. Fixed media filters use microorganisms attached to a medium (rocks, plastic, metal, etc.) The microorganisms stay in place and do not need to be cycled through the system. Instead, wastewater is circulated past the fixed microorganisms. A fixed media filter mimics the treatment method used in a healthy stream in which microorganisms produce a slick coating on rocks and pebbles. This coating of microorganisms is able to trap and consume B.O.D. and ammonia in the water.

The trickling filter consists of several major components including distribution system, media, under drains, effluent channel, secondary settling tank, and recirculation pumps and piping **(Fig 4.5)**. Each of these components has one or more purposes. In operation, wastewater is distributed evenly over the surface of the trickling filter media. As the wastewater flows over the surface of the media the organisms in the slime remove the organic matter from the flow.

TFs enable organic material in the wastewater to be absorbed by a population of microorganisms (aerobic, anaerobic, and facultative bacteria; fungi; algae; and protozoa) attached to the medium as a biological film or slime layer (approximately 0.1 to 0.2 mm thick). As the wastewater flows over the medium, microorganisms already in the water gradually attach themselves to the rock, slag, or plastic surface and form a film. The organic material is then degraded by the aerobic microorganisms in the outer part of the slime layer.

As the layer thickens through microbial growth, oxygen cannot penetrate the medium face, and anaerobic organisms develop. As the biological film continues to grow, the microorganisms near the surface lose their ability to cling to the medium, and a portion of the slime layer falls off the filter. This process is known as sloughing. The sloughed solids are picked up by the under drain system and transported to a clarifier for removal from the wastewater.

Advantages of Trickling Filters

a) Simple, reliable, biological process.

b) Suitable in areas where large tracts of land are not available for land intensive treatment systems.

c) May qualify for equivalent secondary discharge standards.

d) Effective in treating high concentrations of organics depending on the type of medium used.

e) Appropriate for small- to medium-sized communities.

f) Rapidly reduce soluble BOD in applied wastewater.

g) Efficient nitrification units.

h) Durable process elements.

i) Low power requirements.

j) Moderate level of skill and technical expertise needed to manage and operate the system.

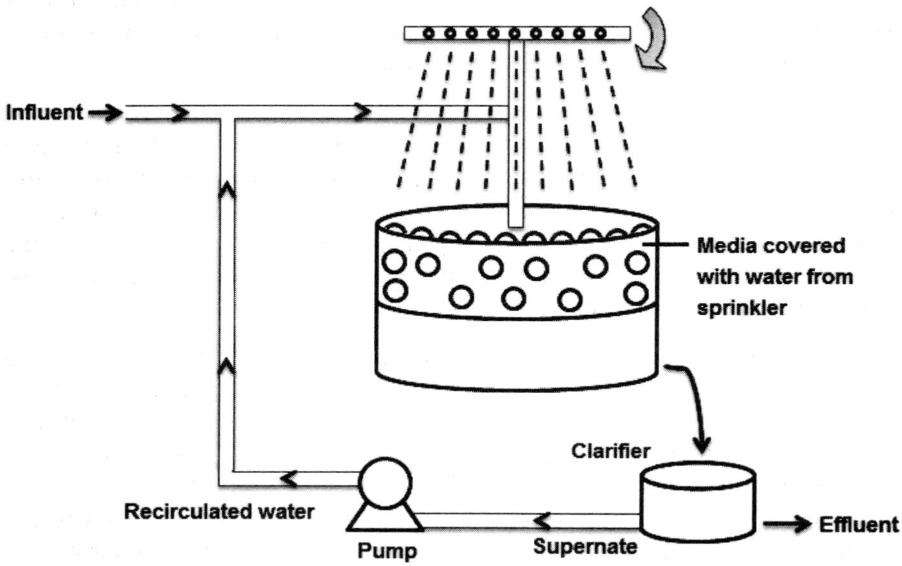

Fig. 4.5: Trickling Filter

Disadvantages of Trickling Filters

a) Additional treatment may be needed to meet more stringent discharge standards.

b) Possible accumulation of excess biomass that cannot retain an aerobic condition and can impair TF performance.

c) Requires regular operator attention.

d) Incidence of clogging is relatively high.

e) Requires low loadings depending on the medium.

f) Flexibility and control are limited in comparison with activated-sludge processes.

g) Vector and odor problems.

Rotating Biological Contactors

A rotating Biological Contactor (RBC) is a biological treatment process used in the treatment of wastewater following primary treatment. The primary treatment process removes the grit and other solids through a screening process followed by a period of settlement. The RBC process involves allowing the wastewater to come in contact with a biological medium in order to remove pollutants in the wastewater before discharge of the treated wastewater to the environment, usually a body of water (river, lake or ocean). A rotating biological contactor is a type of secondary treatment process. It consists of a series of closely spaced, parallel discs mounted on a rotating shaft which is supported just above the surface of the waste water. Microorganisms grow on the surface of the discs where biological degradation of the wastewater pollutants takes place **(Fig 4.6)**.

A rotating biological contractor (RBC) is an attached-growth biological process that consists of one or more basins in which large closely-spaced circular disks mounted on horizontal shafts rotate slowly through waste-water. The disks, which are made of high-density polystyrene or polyvinyl chloride (PVC), are partially submerged in the wastewater, so that a bacterial slime layer forms on their wetted surfaces. As the disks rotate, the bacteria are exposed alternately to waste-water, from which they adsorb organic matter, and to air, from which they absorb oxygen. The rotary movement also allows excess bacteria to be removed from the surfaces of the disks and maintains a suspension of sloughed biological solids. A final clarifier is needed to remove sloughed solids. Organic matter is degraded by means of mechanisms similar to those operating in the trickling filters process. Partially submerged RBCs are used for carbonaceous BOD removal, combined carbon oxidation and nitrification, and nitrification of secondary effluents. Completely submerged RBCs are used for de-nitrification.

Advantages offered by RBCs include:

a) Short contact periods are required because of the large active surface.

b) RBCs are capable of handling a wide range of flows.

c) Sloughed biomass generally has good settling characteristics and can easily be separated from the waste stream.

d) Operating costs are low because little skill is required in plant operation.

e) Short retention time.

f) Low power requirements.

g) Elimination of the channeling to which conventional percolators are susceptible.

h) Low sludge production and excellent process control.

Disadvantages of RBCs include:

a) Requirement for covering RBC units in northern climates to protect against freezing.

b) Shaft bearings and mechanical drive units require frequent maintenance.

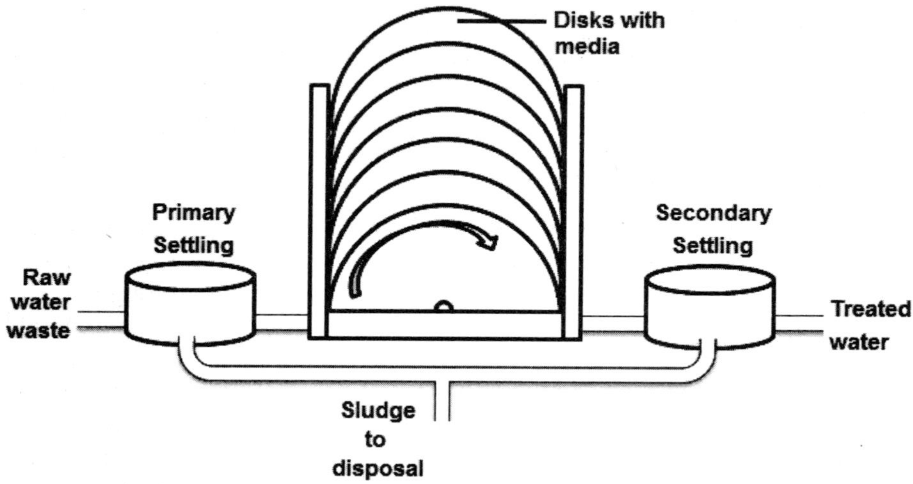

Fig. 4.6: Rotating Biological Contractor

Activated Sludge Method

The term activated sludge refers to suspended aerobic sludge consisting of flocs of active bacteria, which consume and remove aerobically biodegradable organic substances from screened or screened and pre-settled wastewater. Activated sludge systems can treat fecal sludge and industrial wastewater as long as the pollutants to be treated are biodegradable.

Activated sludge reactors are aerobic suspended-growth type processes (in opposition to fixed-film or attached-growth processes **(Fig 4.7)**. Large amounts of injected oxygen allow maintaining aerobic conditions and optimally mixing the active biomass with the wastewater to be treated. To maintain a relatively high amount of active microorganisms useful in removing organic substances from the wastewater, the sludge is separated from the effluent by settling in a secondary clarifier or by membrane filtration and kept in the process by recirculation to the aeration tank.

Several modifications of this basic process have been developed, including different aeration devices, different means of sludge collection and recycling to the aeration tank or primary clarifier, and process enhancement trough the addition of an inert media area on which biofilms can grow (combined fixed-film/suspended-growth process). Although aerobic bacteria are the most dominant microorganisms in the process, other aerobic, anaerobic and/or nitrifying bacteria along with higher organisms can be present. Thus, besides the removal of organic matter, nutrients (organic ammonia, phosphorus) can also be removed biologically by nitrification/denitrification and biological uptake of phosphorus. The exact composition of microorganisms depends on the reactor design, the environment and the wastewater characteristics. To achieve optimal conditions for both, organic and nutrients removal, a sequences of changing aerobic and anaerobic chambers are used.

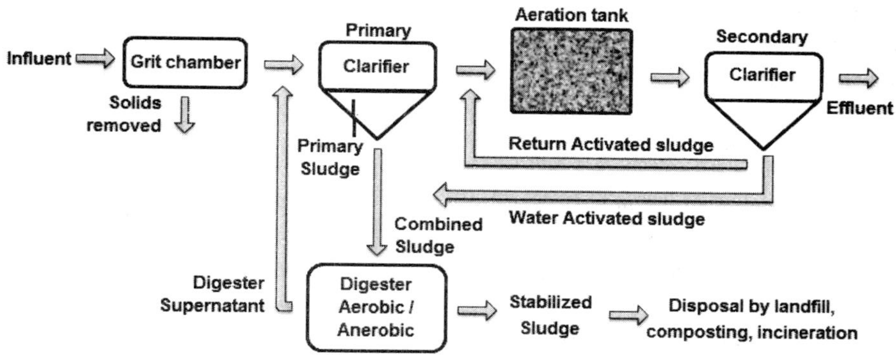

Fig. 4.7: Activated sludge reactor

There are a number of factors that affect the performance of an activated sludge treatment system. These include:

a) Temperature

b) Return rates

c) Amount of oxygen available

d) Amount of organic matter available

e) pH

f) Waste rates

g) Aeration time

h) Wastewater toxicity

To obtain desired level of performance in an activated sludge system, a proper balance must be maintained between the amounts of food (organic matter), organisms (activated sludge) and oxygen (dissolved oxygen).

Activated sludge systems are highly efficient for organic matter and nutrient removal, though pathogen removal is low. In the view of reuse of the effluent in agriculture, it is not beneficial to remove all nutrients while standards for pathogen removal are barely met. As treatment occurs by biological processes, activated sludge could be considered as a naturally based technology. Yet, it does not fit the definition entirely because of the need for high and ongoing energy inputs that make the technology expensive to operate and maintain. As the system is also of high complexity and strongly mechanized, it is mainly adapted for centralized systems where energy, mechanical and technical spare equipment and skilled staff are available.

Ponds

Ponds are probably one of nature's most economical ways of treating sewage and producing a highly purified effluent (end product.) The degree of treatment provided by ponds depends upon the type and number of ponds used. Ponds can be used as the sole type of water treatment or can be used in conjunction with other forms of wastewater treatment.

Advantages and Disadvantages

Ponds have many advantages and disadvantages compared to treatment in plants. Both have to deal with aeration of the water being treated, but in ponds, oxygen is transferred directly into the water across the surface area without the need for any equipment. A plant, in contrast, must install an aerator to add oxygen to the water.

Types of Ponds by Location

a) Raw Sewage Stabilization Pond

b) Oxidation Pond

c) Polishing Pond

These three types of ponds can be used in a series. Alternatively, they may be used in conjunction with primary, secondary, and tertiary treatment in a wastewater treatment plant. The wastewater may receive primary treatment in the treatment plant then receive secondary treatment in an oxidation pond. Or the wastewater may receive primary and secondary treatment in a treatment plant and tertiary treatment in a polishing pond.

Raw Sewage Stabilization Pond

The raw sewage stabilization pond is the most common type of pond. It is a primary treatment facility which receives wastewater which has had no prior treatment (except screening or shredding).

Like any other primary treatment facility, the purpose of the raw sewage stabilization pond is to settle out most of the solids in the water. In addition, aerobic, facultative, and anaerobic decomposition of organic matter begins in this pond **(Fig 4.8)**. Oxygen is provided by diffusion from the surface of the pond and from photosynthesis by the algae in the pond. All of these processes occur over the minimum 45 day detention time during which the water stays in the stabilization pond. The stabilization pond consists of an influent structure, berms or walls surrounding the pond, and an effluent structure designed to permit selection of the best quality effluent. The normal operating depth of the pond is 3 to 5 feet.

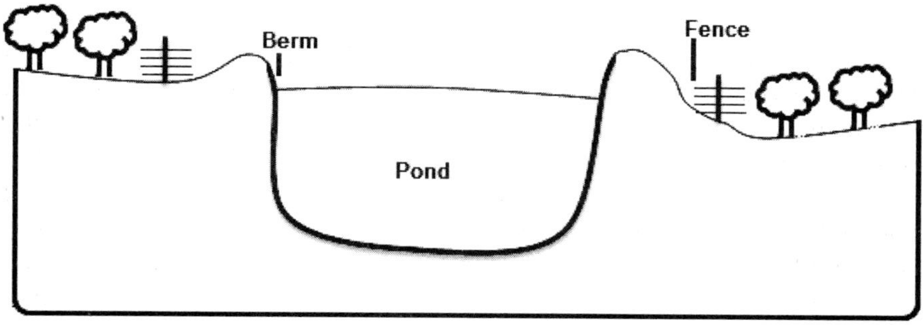

Fig. 4.8: Raw sewage stabilization pond

Oxidation Pond

Water from the stabilization pond or from primary settling tanks of a treatment plant flows into the oxidation pond. In this pond, additional settling of solids and biological treatment of organic matter in the water occurs. Some of the fecal coli form in the water is also removed. The oxidation pond is very similar in design to the stabilization pond.

Polishing Pond

These ponds, also known as finishing ponds, receive water flowing from the oxidation pond or from some other secondary treatment systems. Here, additional BOD, solids, fecal coli form, and some nutrients are removed from the water. Polishing ponds have a much shorter detention time than stabilization ponds since they rely entirely on biological processes and no settling occurs here. Water remains in polishing ponds for only 1 to 3 days. A greater detention time may result in an increased concentration of suspended solids in the effluent. In addition, polishing ponds are typically deeper than the other types of ponds, usually operating at a depth of 5 to 10 feet.

One of the largest problems when water is released directly from a sewage pond into a stream is algae. Sewage ponds are perfect environments for these one-celled plants. Food is readily available, as is moisture and sunlight, so algae grow quickly and become quite numerous.

When water from a sewage pond, rich in algae, is released directly into a stream the stream can be harmed. The large quantities of algae consume oxygen in water at night or during an algal bloom (when the algae reproduce very quickly). Without the oxygen they need to survive, the fish in the stream die. A finishing pond, also known as a polishing pond, can eliminate this problem. Finishing ponds are usually stocked with fish, such as carp, which eat the algae in the water. Finishing ponds also allow the quality of the effluent to be monitored before it is released into the stream. As a result, streams being fed by finishing ponds tend to be healthier than those fed directly from sewage ponds.

Anaerobic Methods of Waste Water Treatment

Although treatment by aerobic bacteria is conducted at a high speed, the supply of oxygen (usually air) is indispensable. Electrical expenses is provided under administrative and maintenance expense. Likewise, the drawback of the treatment method is the mass generation of sludge. Further expenses remain for the industry to deal with. On the other hand, in the case of anaerobic treatment using bacteria, the overall activity is slow, the entire duration of the treatment process is longer, and the rate of removal of organic matter is low, as well. However, the benefit of the system is the minimal generation of sludge only. As a result, the device for the increase of processing efficiency has so far been tried tremendously. Thus, one point of view being considered is the combination of aerobic and anaerobic treatment processes.

It is a general fact that anaerobic biochemical treatment by using anaerobic bacteria is better for high-concentration organic effluent. A familiar example is that human excrement and sewer sludge, which are highly organic in concentration, are treated anaerobic. Whereas municipal wastewater containing lower organic content, is treated aerobically. It is the usual way of treating common wastewater. The decomposition reaction of organic matter, which is conducted under oxygen-free environment, is quite complicated as compared with the treatment system under the presence of oxygen. The primary feature is the production of many kinds of substances. The main products are methane and carbon dioxide. Other by-products include ammonia, hydrogen, and hydrogen sulfide. Especially, since natural fuel sources, such as methane and hydrogen, are generated, the viewpoint of bio-energy comes in.

Anaerobic treatment is based on microbiological processes, namely methane fermentation, which occurs in an anaerobic environment. Numerous species of bacteria have to cooperate in order to convert the organic pollution in the water to a mixture of methane (CH_4) and carbon dioxide (CO_2), called biogas. The bacteria are generally present as sludge flocs or bacterial clusters (aggregates).

This process has been traditionally more complex and consequently harder to control than the aerobic biological process used in the classic activated sludge wastewater treatment. Better understanding of the microbiology of anaerobic processes has resulted in the successful development of new, improved and practical systems. Anaerobic digestion generally has following steps **(Fig 4.9)**:

Hydrolysis and Acidogenesis

Complex particulate and solubilized polymeric substrates (e.g. polysaccharides and proteins) are hydrolysed to simpler soluble mole-cules (amino acids and sugars). These products are then further catabolized by fermentative micro-organisms, to produce mainly volatile fatty acids (VFA), aldehydes, alcohols, carbon dioxide and hydrogen.

Acetogenesis

The majority of the fermentation products, except H_2, CO_2, formate and acetate, are further degraded by the -acetogens to yield acetate and H_2 and additional CO_2. The acetogens grow in close association with the -methanogenic bacteria.

Methanogenesis

The final step in the anaerobic digestion is carried out by the methanogenic bacteria and is the formation of methane- gas from acetate and from hydrogen and carbon dioxide.

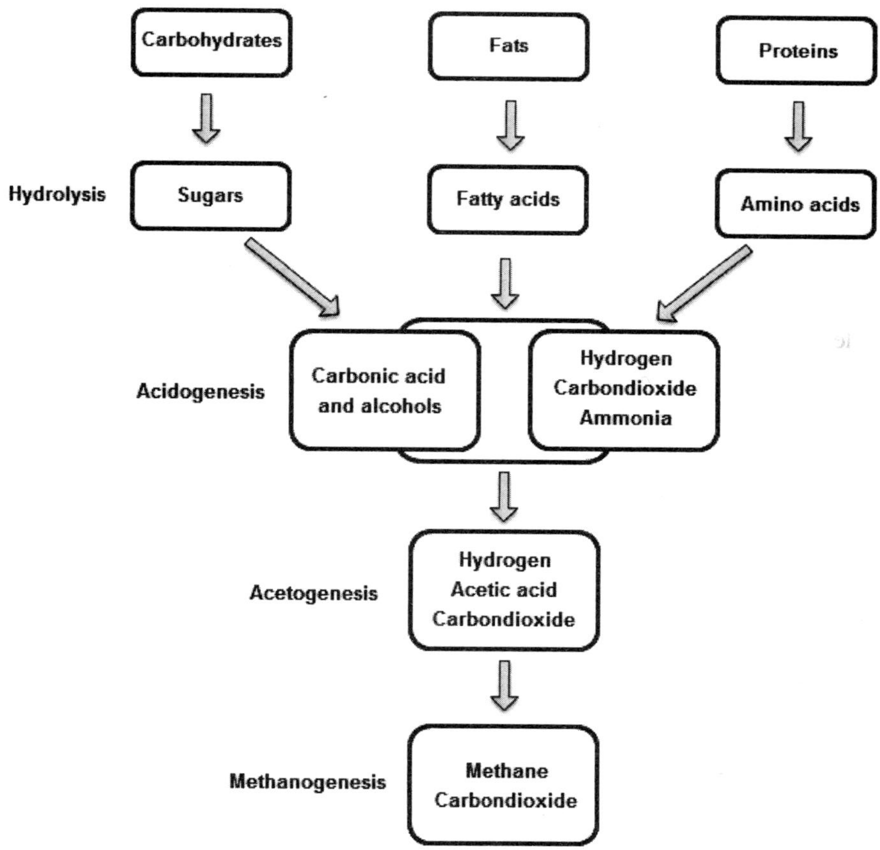

Fig. 4.9: Steps of anaerobic digestion

Types of Anaerobic Reactors

There are four principal process variants which are proper in anaerobic wastewater treatment. These are as follows:

a) **Anaerobic Filter**: The anaerobic filter is similar to a trickling filter in that a biofilm is generated on media. The bed is fully submerged and can be operated either upflow or downflow. For very high strength wastewaters, a recycle can be employed.

b) **Anaerobic Contact**: This process can be considered as an anaerobic activated sludge because sludge is recycled from a clarifier or separator to the reactor. Since the material leaving the reactor is a gas-liquid-solid mixture, a vacuum degasifier is required to separate the gas and avoid floating sludge in the clarifier.

c) **Fluidized Bed**: This reactor consists of a sand bed on which the biomass is grown. Since the sand particles are small, a very large biomass can be developed in a small volume of reactor. In order to fluidize the bed, a high recycle is required.

d) **Upflow Anaerobic Sludge Blanket (UASB)**: Under proper conditions anaerobic sludge will develop as high density granules. These will form a sludge blanket in the reactor. The wastewater is passed upward through the blanket. Because of its density, a high concentration of biomass can be developed in the blanket.

Anaerobic Filter

The anaerobic filter, also known as fixed bed or fixed film reactor is used for the treatment of non-settelable and dissolved solids by bringing them in close contact with a surplus of active bacterial mass. This surplus together with "hungry" bacteria digests the dispersed or dissolved organic matter within short retention times. Anaerobic filters are reactors consisting of supporting material layers. On the surface of these material layers or bed, fixation of microorganism and the development of biofilm take place. Anaerobic filters can be applied not only for treating concentrated wastewater but also for those wastewaters that have low organic load (grey water). If they are preceded by a reactor that retains settled solids, they will work better. It is suitable for domestic wastewater and all industrial wastewater which have a lower content of suspended solids. The bacteria in the filter are immobile and generally attach themselves to solid particles or to the reactor walls. Filter materials like rocks, cinder, plastic, or gravel provide additional surface area for bacteria to settle. Thus, the fresh wastewater is forced to come into contact with active bacteria intensively. The larger surface area for the bacterial growth helps in the quick digestion of the wastes. A good filter material provides a surface area of 90 to 300 m^2per meter cube reactor volume. Biological oxygen demand up to 70% to 90 % is removed in a well operated anaerobic filter.

Pre-treatment in settlers or septic tanks may be necessary to eliminate solids of larger size before they are allowed to enter the filter. When the bacterial film becomes too thick it has to be removed. This may be done by back-flush of wastewater or by removing the filter mass for cleaning outside the reactor. Nonetheless, the anaerobic filter is very reliable and robust. Anaerobic filters may be operated as down flow or up flow systems **(Fig 4.10)**. A combination of up-flow and down-flow chambers is also possible.

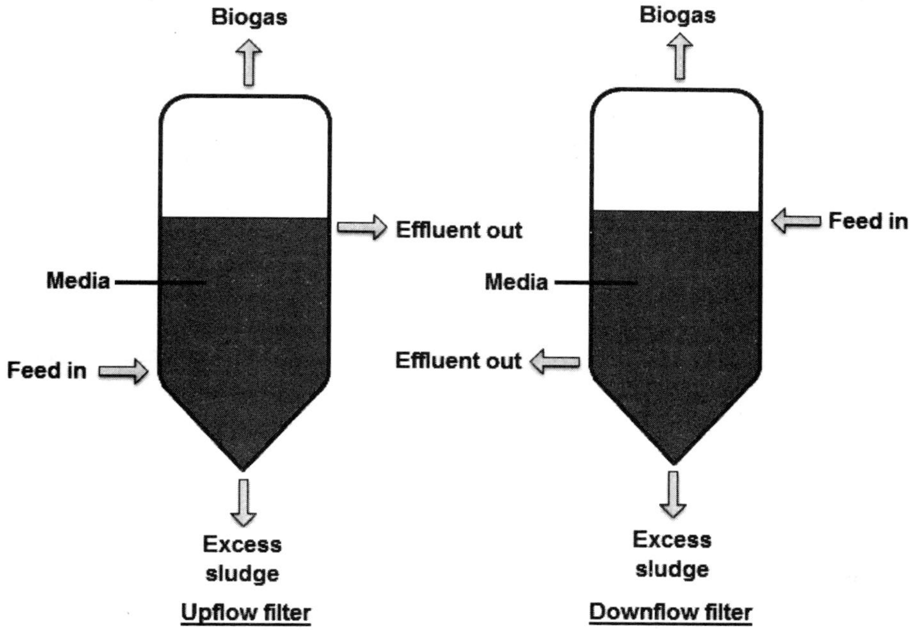

Fig. 4.10: Upflow and Downflow anaerobic filter

Anaerobic Contact Process

Anaerobic contact reactors employ an external clarifier or vessel to settle solids and subsequently recycle them back to the reactor tank. Typical configurations include large tanks due to the low organic and hydraulic loading rates employed in their design **(Fig 4.11)**. Anaerobic contact systems are particularly effective when granulation is difficult or wastewater contains higher than desirable amounts of troublesome constituents, e.g. O&G, suspended solids. Anaerobic contact alternatives are effective at successfully retaining flocculent, i.e. non-granular sludge, thus permitting maintaining appropriate anaerobic biomass inventory levels.

Fluidized Bed Reactor

FBR is a biological reactor that accumulates a maximum active attached biomass yet still handling fine suspended solids without blocking. By maximizing the surface area available for microbial attachment and minimizing the volume occupied by the media, a maximum specific activity of attached biomass may be achieved for a given reactor volume. A filter containing extremely small particles (0.5 mm) provides adequate surface area to achieve these benefits. In order to achieve fluidization of the biomass particles, units must be operated in an upflow mode. Rate of liquid flow and the resulting degree of bed expansion determines whether the reactor is termed a fluidized bed or expanded bed system. Expanded bed reactors have a bed expansion of 10% to 20% compared to 30% to 90% in fluidized beds.

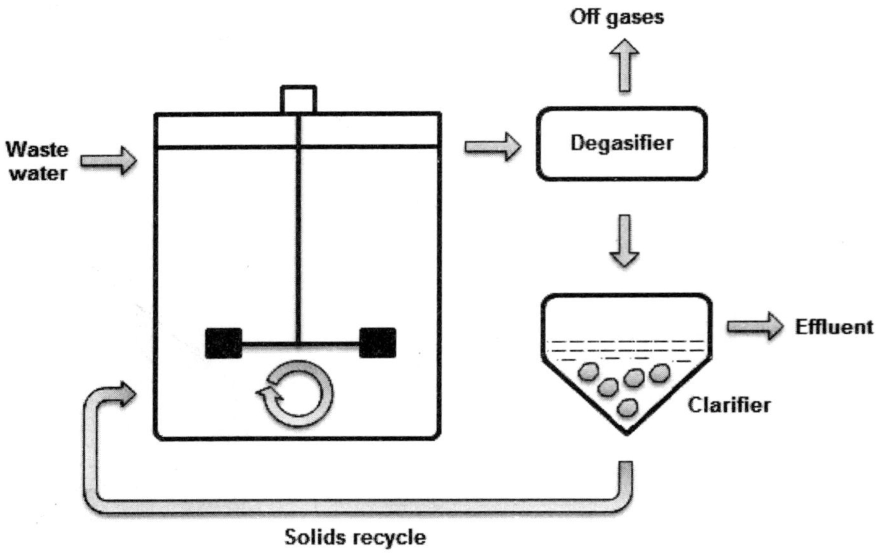

Fig. 4.11: Anaerobic contact process

In FBR, biomass is attached to surface of small particles like anthracite, high density plastic beads, sand etc. **(Fig 4.12)** which are kept in suspension by upward velocity of liquid flow. Effluent is recycled to dilute incoming waste and to provide sufficient flow-rate to keep particles in suspension. Large surface area of support particles and high degree of mixing that result from high vertical flows enable a high biomass concentration to develop and efficient substrate uptake.

The greatest risk with FBR is the loss of biomass particles from the reactor following sudden changes in particle density, flow rate or gas production. If flow is interrupted and the bed allowed settling, there is a tendency once flow is restarted for the entire bed to move upward in plug-flow rather than fluidizing. In practice, considerable difficulties were experienced in controlling the particle size and density of flocs due to variable amounts of biomass growth on particles. Therefore FBRs are considered to be difficult to operate.

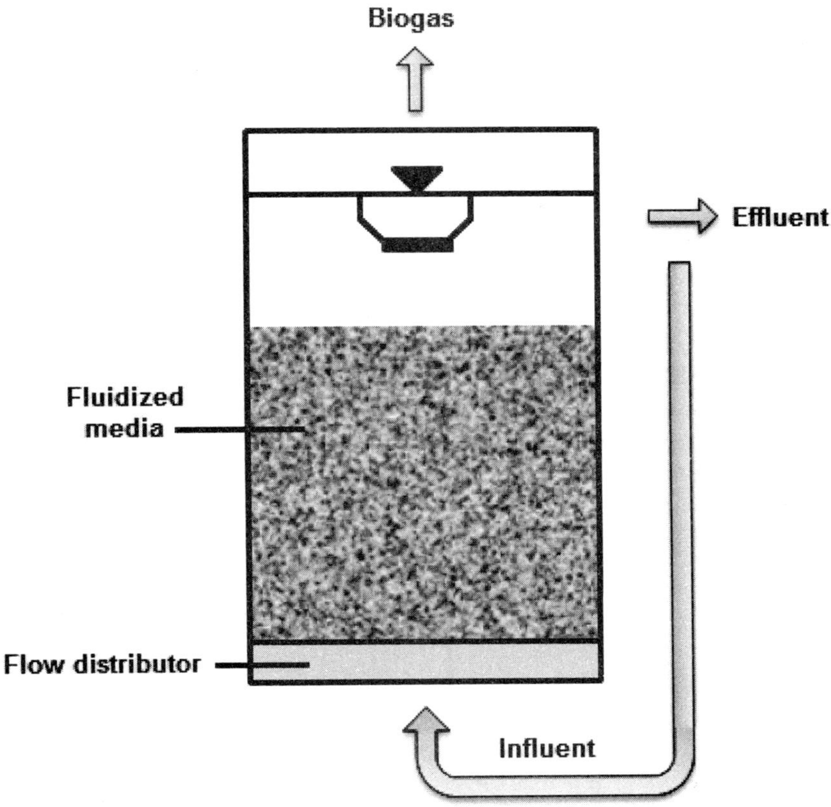

Fig. 4.12: Fluidized bed reactor

Up flow Anaerobic Sludge Blanket Reactor

Up flow anaerobic sludge blanket technology also known as UASB reactor is a form of anaerobic digester which used in wastewater treatment. UASB reactor is a methane-producing digester, which uses an anaerobic process and forming a blanket of granular sludge and is processed by the anaerobic microorganisms. UASB reactor is based on the so-called three-phase separator, which enables the reactor to separate gas, water and sludge mixtures under high turbulence conditions **(Fig 4.13)**. This allows for compact, cheaper designs.

The reactor has multiple gas hoods for the separation of biogas. As a result the extremely large gas/water interfaces greatly reduce turbulence, making relatively high loading rates of 10 - 15 kg/m^3 possible. Separation in the UASB reactor requires only 1.0 meter of height, which prevents flotation effects and, consequently, floating layers.

Generally, during the treatment of UASB reactor, the substrate passes through an expanded sludge bed which containing a high concentration of biomass first. After that, the remaining part of substrate passes through a less dense biomass which named the sludge blanket. The influent is pumped to the UASB reactor from bottom of it by Peristaltic pump. The influent move upwards and get contact with the biomass in sludge bed, then continue to move upwards and the rest substrates act with the biomass again in the sludge blanket which has a less concentration of biomass compared with the sludge bed below.

The volume of sludge blanket must be sufficient to conduct the further treatment to wastewater by-passed from the lower layer of sludge bed by channeling. At the same time, it will help to ensure a stable effluent quality. A 3 phases (Gas-Liquid-Solid or GLS) separator located above the sludge blanket to separate the solid particles from the mixture (gas, liquid, and solid) after treatment and hence allowing liquid and gas to leave the UASB reactor. After the treated wastewater will be collected by the effluent collection system via number of launders distributed over entire area discharging, to main launder provided at periphery of the reactor. And the biogases generated will be collected as the valuable fuel or for deposal.

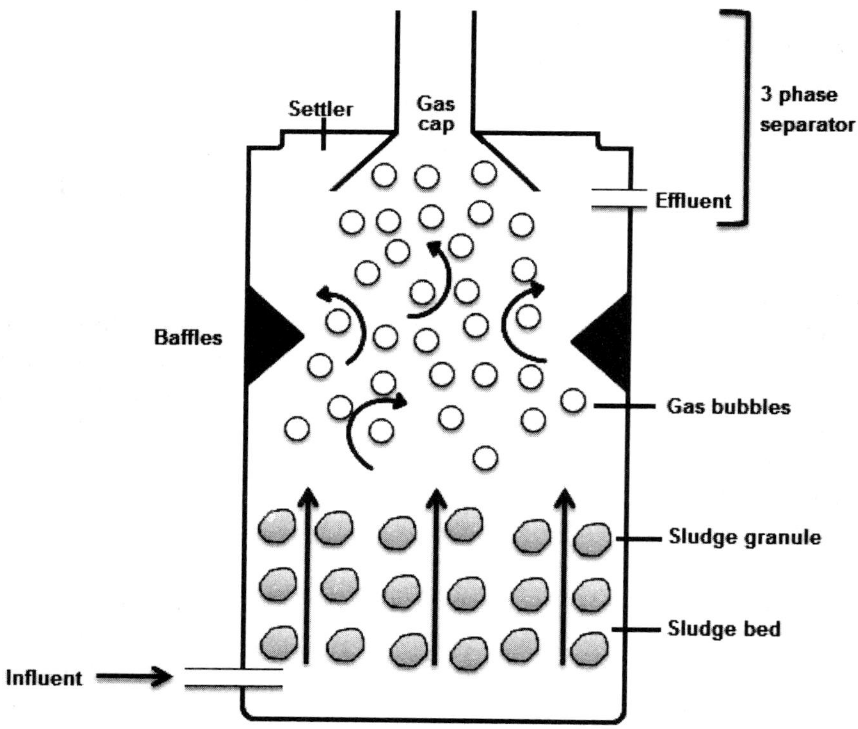

Fig. 4.13: Upflow anaerobic sludge blanket reactor

Membrane Bioreactors

MBR technology is based on the combination of conventional activated sludge treatment together with a process filtration through a membrane with a pore size between 10 nm and 0.4 microns (micro/ultrafiltration), which allows sludge separation. The membrane is a barrier that retains all particles, colloids, bacteria and viruses, providing a complete disinfection of treated water. Furthermore, it can operate at higher concentrations of sludge (up to 12 g/l instead of the usual 4 g/l in conventional systems), which significantly reduces the volume of the reactors and sludge production.

When used with domestic wastewater, MBR processes can produce effluent of high quality enough to be discharged to coastal, surface or brackish waterways or to be reclaimed for urban irrigation. Other advantages of MBRs over conventional processes include small footprint, easy retrofit and upgrade of old wastewater treatment plants. When used with domestic wastewater, MBR processes can produce effluent of high quality enough to be discharged to coastal, surface or brackish waterways or to be reclaimed for urban irrigation. Other advantages of MBRs over conventional processes include small footprint, easy retrofit and upgrade of old wastewater treatment plants.

During MBR wastewater treatment, solid–liquid separation is achieved by Microfiltration (MF) or Ultrafiltration (UF) membranes. A membrane is simply a two-dimensional material used to separate components of fluids usually on the basis of their relative size or electrical charge. The capability of a membrane to allow transport of only specific compounds is called semi-permeability. This is a physical process, where separated components remain chemically unchanged. Components that pass through membrane pores are called permeate, while rejected ones form concentrate or retentate. There are five types of membrane configuration which are currently in operation:

a) Hollow fibre (HF)

b) Spiral-wound

c) Plate-and-frame (i.e. flat sheet (FS))

d) Pleated filter cartridge

e) Tubular

Types of MBR

Internal/submerged

The filtration element is installed in either the main bioreactor vessel or in a separate tank. The membranes can be flat sheet or tubular or combination of both, and can incorporate an online backwash system which reduces membrane surface fouling by pumping membrane permeate back through the membrane **(Fig 4.14)**.

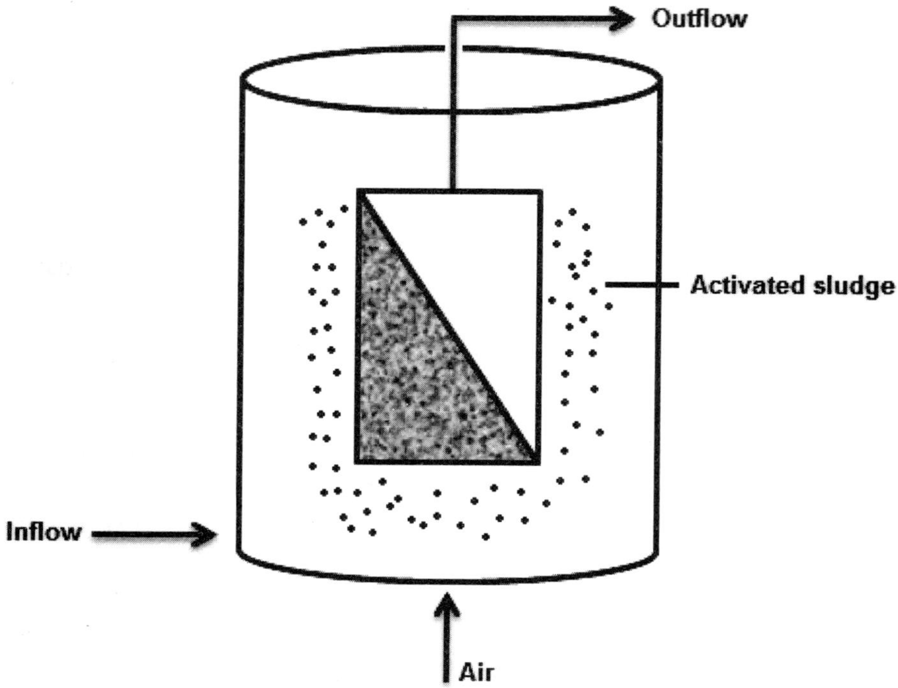

Fig. 4.14: Submerged membrane bioreactor

External/side stream

The filtration elements are installed externally to the reactor, often in a plant room. The biomass is either pumped directly through a number of membrane modules in series and back to the bioreactor, or the biomass is pumped to a bank of modules, from which a second pump circulates the biomass through the modules in series **(Fig 4.15)**. Cleaning and soaking of the membranes can be undertaken in place with use of an installed cleaning tank, pump and pipework.

Advantages

a) The effluent is of very high quality, very low in BOD (less than 5 mg/l), very low in turbidity and suspended solids. The technology produces some of the most predictable water quality known. It is fairly easy to operate as long as the operation has been properly trained, pays strict attention to the proper operation, corrective maintenance, and preventative maintenance tasks.

b) The "simple filtering action" of the membranes creates a physical disinfection barrier, which significantly reduces the disinfection requirements.

c) The capitol cost is usually less than for comparable treatment trains.

d) The treatment process also allows for a smaller "footprint" as there are neither secondary clarifiers nor tertiary filters which would be required to achieve similar water quality results. It also eliminates the need for a tertiary backwash surge tank, a backwash water storage tank, and for the treatment of the backwash water.

e) Generally speaking it produces less waste activated sludge than a simple conventional system.

f) If reuse is a major water quality goal, the MBR process will be a major consideration. This process produces a consistent, high water quality discharge. When followed by a disinfection process, it allows for a wide range of water re-use applications including landscape irrigation, non-root edible crops, highway median strip and golf course irrigation, and cooling water re-charge. When Reverse Osmosis (RO) water quality is required, the MBR process is an excellent candidate for preparing the water for RO treatment.

Disadvantages

a) The membrane modules will need to be replaced somewhere between five (5) and ten (10) years with the current technology. While the costs have decreased over the past several years, these modules can still be classified as expensive.

b) Fouling is troublesome, and its prevention is costly. Several papers and research endeavors have concluded that up to two-thirds of the chemical and energy costs in an MBR facility are directly attributable to reducing membrane fouling.

c) There may be cleaning solutions that require special handling, treatment, and disposal activities depending on the manufacturer. These cleaning solutions may be classified as hazardous waste depending on local and state regulations.

Advance Waste Water Treatment/Tertiary Treatment

Advanced treatment is the treatment to further remove nitrogen, phosphate, solids, salt, color and odor from the biologically treated effluent as well as to achieve disinfection. Advanced treatment is also called "Tertiary Treatment". The purpose of tertiary treatment is to provide a final treatment stage to raise the effluent quality to the desired level. This advanced treatment can be accomplished by a variety of methods such as coagulation sedimentation, filtration, reverse osmosis, and extending secondary biological treatment to further stabilize oxygen-demanding substances or remove nutrients. In various combinations, these processes can achieve any degree of pollution control desired. As wastewater is purified to higher and higher degrees by such advanced treatment processes, the treated effluent can then be reused for urban, landscape, and agricultural irrigation, industrial cooling and processing, recreational uses and water recharge, and even indirect and direct augmentation of drinking water supplies.

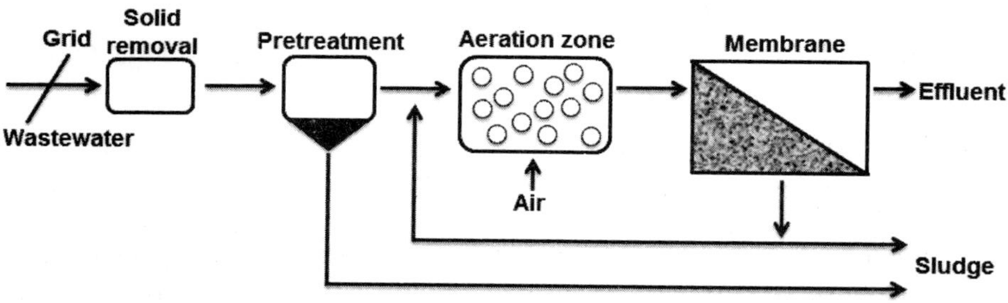

Fig. 4.15: External membrane bioreactor

Coagulation Sedimentation

Chemical coagulation sedimentation is used to increase the removal of solids from effluent after primary and secondary treatment. Solids heavier than water settle out of wastewater by gravity. With the addition of specific chemicals, solids can become heavier than water and will settle. Alum, lime, or iron salts are chemicals added to the wastewater to remove phosphorus. With the chemicals, the smaller particles clump or 'flocs' together into large masses. The larger masses of particles will settle out in the sedimentation tank reducing the concentration of phosphorus by more than 95%.

Filtration

A variety of filtration methods are available to ensure high quality water. Sand filtration, which consists of simply directing the flow of water through a sand bed, is used to remove residual suspended matter. Filtration over activated carbon results in the removal of the following types of contaminants: non-biodegradable organic compounds, absorbable organic halogens, toxins, color compounds and dyestuffs, aromatic compounds including phenol and bis-phenol A (BPA), chlorinated/halogenated organic compounds, and pesticides. Although there are a number of different methods of membrane filtration, the most mature is pressure driven membrane filtration. This relies on a liquid being forced through a filter membrane with a high surface area. Membrane filtration is designed to remove bacteria, viruses, pathogens, metals, and suspended solids.

Reverse Osmosis

In the reverse osmosis process, pressure is used to force effluent through a membrane that retains contaminants on one side and allows the clean water to pass to the other side. Reverse osmosis is actually a type of membrane filtration called microfiltration because it is capable of removing much smaller particles including dissolved solids such as salt. This process is also effective at removing biological contaminants, metals, pharmaceuticals, pesticides, and endocrine disruptors.

Nitrogen Removal

To remove nitrogen from wastewater using biological nitrification/de-nitrification, chlorination, ion exchange, etc.

Phosphate Removal

To remove phosphates from wastewater using biological phosphorus removal, chemical precipitation and combining biological and chemical processes.

Solid (Colloids) Removal

It is a method to remove suspended and dissolved solids. Suspended solids, including fine particles and colloids cannot be removed efficiently through gravity sedimentation; they are removed using centrifuge, dissolved air flotation (DAF), diatomaceous earth filtration, coagulation sedimentation plus rapid sand filtration; dissolved solids are removed with ion exchange, membranes, activated carbon adsorption and chemical oxidation.

The overall philosophy of wastewater sanitation involving the removal, control and treatment of a wastewater in an area that is isolated or remote from the center of activity is important. Over the years wastewater treatment management practices have evolved into a technically complex body of knowledge based on past practice and applied engineering and environmental sciences. The intelligent application of these fundamentals goes a long way toward assuring us that the environment will be maintained in a safe and acceptable condition.

REFERENCES

Asano, T.; Burton, F., Leverenz, H.; Tsuchinashi, R. & Tchobanoglous, G. (2006). Water Reuse: Issues, Technologies and Applications. Metcalf & Eddy/AECOM. ISBN: 978-0-07-145927-3. 1st ed.

Ayala, D.F.; Ferre, V. & Judd. S.J. (2011). Membrane life estimation in full-scale immersed membrane bioreactors. Journal of Membrane Science (in press), doi: 10.1016/j.memsci. 2011.03.013.

BCC. (2011). Membrane bioreactors: global markets. BCC Report MST047C. March 2011.

Brepols, C.; Dorgeloh, E.; Frechen, F.-B.; Fuchs, W. ; Haider, S.; Joss, A.; de Korte, K. ;Ruiken, C.; Schier, W.; van der Roest, H.; Wett, M. & Wozniak, T. (2008). Upgrading and retrofitting of municipal wastewater treatment plants by means of membrane bioreactor (MBR) technology. Desalination, Vol. 231, No. 1-3, pp. 20-26.

Chang, I.S.; Le-Clech, P.; Jefferson, B. & Judd, S. (2002). Membrane fouling in membrane bioreactors for wastewater treatment. Journal of Environmental Engineering, Vol. 128, No. 11, pp. 1018–1029.

Cicek, N., Franco, J.P., Suidan, M.T., Urbain, V., Manem, J. (1999). Characterization and comparison of a membrane bioreactor and a conventional activated-sludge system in the treatment of wastewater containing high-molecular-weight compounds. Water Environ. Res., Vol. 71, No. 1, pp. 64-70.

Coello Oviedo, M.D., López-Ramírez, J.A., Sales Márquez, D. & Quiroga Alonso, J.M. (2003). Evolution of an activated sludge system under starvation conditions. Chem. Eng. J.,Vol. 94, pp. 139-146.

Cui, Z.F., Chang, S. & Fane, A.G. (2003). The use of gas bubbling to enhance membrane processes, Journal of Membrane Science, Vol. 221, pp. 1–35.

Defrance L., Jaffrin, M.Y.; Gupta, B.; Paullier, P. & Geaugey, V. (2000). Contribution of various constituents on activated sludge to membrane bioreactor fouling. Bioresource Technology, Vol. 73, pp. 105-112.

Delgado, S.; Villarroel, R.; González, E. (2010). Submerged Membrane Bioreactor at Substrate-Limited Conditions: Activity and Biomass Characteristics. Water Environment Research, Vol. 82, No. 3, pp. 202-208.

Di Bella G., Durante, F., Torregrossa, M. & Viviani, G. (2006). The role of fouling mechanisms in submerged membrane bioreactor during the start-up. Desalination, Vol. 200, pp. 722-724.

Drews, A. (2010). Membrane fouling in membrane bioreactors—Characterisation, contradictions, cause and cures. Journal of Membrane Science, Vol. 363, No. 1-2, pp. 1-28.

Dubois, M.; Gilles, K.A.; Hamilton, J.K.; Rebers, P.A. & Smith, F. (1956). Calorimetric method for determination of sugars and related substances. Anal Chem., Vol. 28, No. 3, pp. 350-356.

Ferrero, G.; Monclús, H.; Buttiglieri, G.; Comas, J. & Rodriguez-Roda, I. (2011). Automatic control system for energy optimization in membrane bioreactors. Desalination, Vol. 268, No. 1-3, pp. 276-280.

Foley, G. (2006) A review of factors affecting filter cake properties in dead-end microfiltration of microbial suspensions. Journal of Membrane Science, Vol. 274, pp. 38–46.

Gawande, N. A., Reinhart, D. R., Thomas, P. A., McCreanor, P. T., and Townsend, T. G. (2003). "Municipal solid waste in situ moisture content measurement using an electrical resistance sensor." Waste Manage., 23(7), 667-674.

Gujer, W. and Zehnder, A.J.B. "Conversion processes in anaerobic digestion." Water Science and Technology, Vol. 15, pp. 127-267, 1983.

Gurijala, K. R., and Suflita, J. M. (1993). "Environmental factors influencing methanogenesis from refuse in landfill samples." Environ. Sci. Technol., 27(6), 1176-1181.

Henze, M.; Harremoes, P. "Anaerobic treatment of wastewater in fixed film reactors – a literature review." Water Science and Technology, Vol. 15, pp. 1-101, 1983.

Hulshoff Pol, L.W.; Lopes, C.I.S.; Lettinga, G.; Lens, L.N.P. "Anaerobic sludge granulation." Water Research??, Vol. 38, pp. 1376-1389, 2004.

Intergovernmental Panel on Climate Change (IPCC) Fourth Assessment Report: Climate Change 2007.

Khanal, Samir Kumar. Anaerobic Biotechnology for Bioenergy Production. Wiley-Blackwell, 2008.

Lettinga, G., Rebac, S., & Zeeman, G. "Challenges of psychrophilic anaerobic wastewater treatment," Trends in Biotechnology, Vol. 19 No. 9, pp. 363-370, 2001.

McCarty, P.L. and Smith, D.P. "Anaerobic wastewater treatment: Fourth of a six-part series on wastewater treatment processes." Environmental Science and Technology, Vol. 20,No. 12, pp. 1200-1206, 1986.

Mehta, R., Barlaz, M. A., Yazdani, R., Augenstein, D., and Bryars, M. (2002). "Refuse Decomposition in the Presence and Absence of Leachate Recirculation." Journal of Environmental Engineering, 128(3), 228-236.

Metcalf & Eddy, Inc. Wastewater Engineering: Treatment and Reuse. Fourth Edition, revised by George Tchobanoglous, Franklin L. Burton, and H. David Stensel. McGraw Hill, 2004.

National Academy of Sciences (NAS). Methane Generation from Human, Animal, and Agricultural Wastes. 1977.

Novaes, R.F.V. Microbiology of anaerobic digestion, Water Science and Technology, Vol. 18, No. 12, pp. 1-14, 1986.

Office of the Leading Group for the Popularisation of Biogas (OLGPB) in Sichuan Province, Peoples' Republic of China, A Chinese Biogas Manual 1978.

Sharma, K.R. "Kinetics and Modeling in Anaerobic Processes" in Anaerobic Technology for Bioenergy Production: Principles and Applications by S.K. Khanal, Ames, Iowa: Wiley-Blackwell, 2008.

Tolaymat, T. M., Green, R. B., Hater, G. R., Barlaz, M. A., Black, P., Bronson, D., and Powell, J. (2010). "Evaluation of Landfill Gas Decay Constant for Municipal Solid Waste Landfills Operated as Bioreactors." Journal of the Air and Waste Management Association, 60 91-97.

Vavilin, V. A., Lokshina, L. Y., Jokela, J. P. Y., and Rintala, J. A. (2004). "Modelling solid waste decomposition." Biosource Technological, (94), 69-81.

Wreford, K. A., Atwater, J. W., and Lavkulich, L. M. (2000). "The effects of moisture inputs on landfill gas production and composition and leachate characteristics at the Vancouver Landfill Site at Burns Bog." Waste Management and Research, 18(4), 386-392.

5

Bioremediation

"Remediate" means to solve a problem, and "bio-remediate" means to use biological organisms to solve an environmental problem such as contaminated soil or groundwater. Bioremediation provides a technique for cleaning up pollution by enhancing the same biodegradation processes that occur in nature. Depending on the site and its contaminants, bioremediation may be safer and less expensive than alternative solutions such as incineration or land filling of the contaminated materials. It also has the advantage of treating the contamination in place so that large quantities of soil, sediment or water do not have to be dug up or pumped out of the ground for treatment.

Bioremediation means to use a biological remedy to abate or clean up contamination. Microbes are often used to remedy environmental problems found in soil, water, and sediments. Plants have also been used to assist bioremediation processes. This is called phytoremediation. Biological processes have been used for some inorganic materials, like metals, to lower radioactivity and to remediate organic contaminants. With metal contamination the usual challenge is to accumulate the metal into harvestable plant parts, which must then be disposed of in a hazardous waste landfill before or after incineration to reduce the plant to ash. Two exceptions are mercury and selenium, which can be released as volatile elements directly from plants to atmosphere.

For the process of bioremediation, it is necessary that microorganisms should be healthy and active so that they can perform their duty efficiently. It is not necessary that all the microorganisms detoxify the same contaminants and toxins but for different toxins there are different microorganisms. Bioremediation can take place in two conditions that are aerobic and anaerobic conditions. Some examples of bioremediation related technologies are phytoremediation, bioventing, bioleaching, landfarming, bioreactor, composting, bioaugmentation, rhizofiltration, and biostimulation.

For example, researchers are exploiting large scale application of bioremediation that can affect desert formation, global climate change and the life cycle of materials. Attempts are being made to develop microorganisms that can help reverse desert formation. This work is based on developing biopolymers that retain water and reverse desert formation. *Alcaligens luteus* is being used to produce 'super bioabsorbent', a polysaccharide which is composed of glucose and glucuronic acid. These can absorb and hold more than thousand times of its own weight of water. The toxic waste materials remain in vapor, liquid or solid phases; therefore, bioremediation technology varies accordingly whether the waste material involved is in its natural setting or is removed and transported into a fermenter (bioreactor). On the basis of removal and transportation of wastes for treatment, basically there are two methods: *in situ* bioremediation and *ex situ* bioremediation **(Fig 5.1)**.

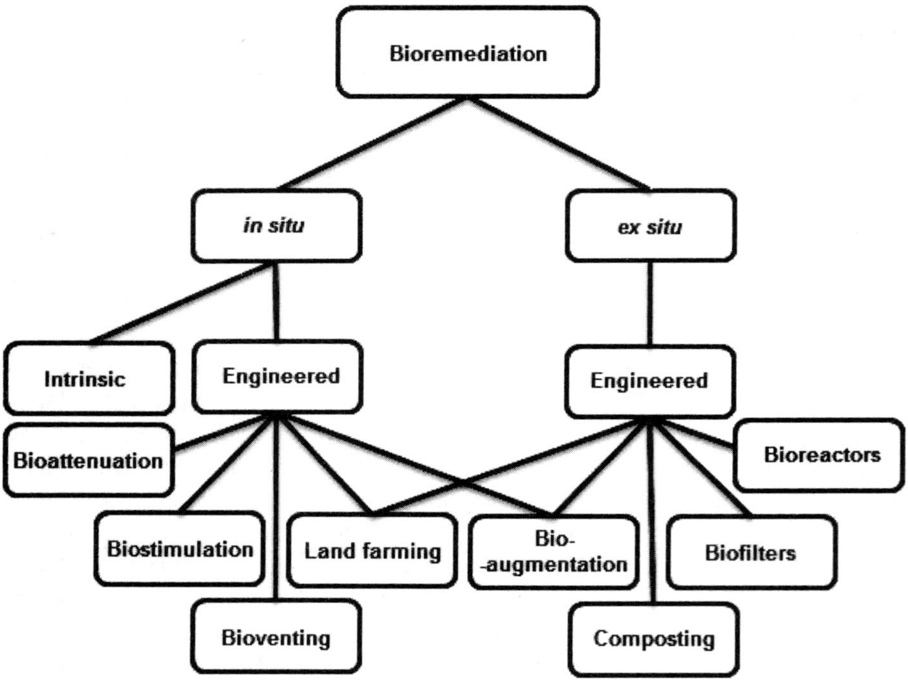

Fig. 5.1: Bioremediation

IN SITU BIOREMEDIATION

Some types of microbes eat and digest contaminants, usually changing them into small amounts of water and harmless gases like carbon dioxide and ethene. If soil and groundwater do not have enough of the right microbes, they can be added in a process called "bioaugmentation." For bioremediation to be effective, the right temperature, nutrients, and food also must be present. Proper conditions allow the right microbes to grow and multiply—and eat more contaminants. If conditions are not right, microbes grow too slowly or die, and contaminants are not cleaned up. Conditions may be improved by adding "amendments." Amendments range from household items like molasses and vegetable oil, to air and chemicals that produce oxygen. Amendments are often pumped underground through wells to treat soil and groundwater *in situ* (in place). In situ bioremediation is the clean up approach which directly involves the contact between microorganisms and the dissolved contaminants for biotransformation.

Potential advantages of in situ bioremediation methods include

1) Minimal site disruption,
2) Simultaneous treatment of contaminated soil and ground water,
3) Minimal exposure of public and site personnel.
4) Low costs.

But the disadvantages are:

1) Time consuming method as compared to other remedial methods,
2) Seasonal variation of microbial activity resulting from direct exposure to prevailing environmental factors, and lack of control of these factors.
3) Problematic application of treatment additives like nutrients, surfactants and oxygen etc.

The microorganisms act well only when the waste materials help them to generate energy and nutrients to build up more cells. When the native microorganisms lack biodegradation capacity, genetically engineered microorganisms (GEMs) may be added to the surface during in situ bioremediation. But stimulation of indigenous microorganisms is preferred over addition of GEMs.

Types of in situ Bioremediation

There are two types of in situ bioremediation: intrinsic and engineered in situ bioremediation.

Intrinsic Bioremediation

In Intrinsic Bioremediation, no external means are employed to speed up remediation. The process is completely governed by natural processes. It is the conversion of environmental pollutants into the harmless forms through the innate capabilities of naturally occurring microbial populations. The conditions of site that favor intrinsic bioremediation are ground water flow throughout the year, carbonate minerals to buffer acidity produced during biodegradation, supply of electron acceptors and nutrients for microbial growth and absence of toxic compounds. The other environmental factors such as pH, concentration, temperature and nutrient availability determine whether or not biotransformation takes place. Bioremediation of waste mixtures containing metals such as Hg, Pb, As and cyanide at toxic concentration can create problem. Natural bioremediation has been occurring for millions of years. Biodegradation of dead vegetation and dead animals is a kind of bioremediation. It is a natural part of the carbon, nitrogen, and sulfur cycles. Chemical energy present in waste materials is used by microorganisms to grow while they convert organic carbon and hydrogen to carbon dioxide and water.

Engineered in situ Bioremediation

When site conditions are not suitable, bioremediation requires construction of engineered systems to supply materials that stimulate microorganisms. Engineered *in situ* bioremediation accelerates the desired biodegradation reactions by encouraging growth of more microorganisms via optimizing physico-chemical conditions. Oxygen and electron acceptors (e.g. NO_3^{1-} and SO_4^{2-}) and nutrients (e.g. nitrogen and phosphorus) promote microbial growth in surface. This may also use genetically engineered microorganism for fast degradation of waste substances in soil and water. The use of genetic engineering to create organisms specifically designed for bioremediation has great potential. For example, the bacterium *Deinococcus radiodurans* (the most radio resistant organism known) has been modified to consume and digest toluene and ionic mercury from highly radioactive nuclear waste.

Some of the strategies utilized for *in situ* bioremediation are as follows:

1) Biostimulation through the addition of chemical additives to enhance or stimulate natural microbial processes;
2) Chemical oxidation through the injection of potassium permanganate, calcium hydroxide, and other reactive slurries into the subsurface;
3) Bioremediation via injection of highly oxygenated water or oxygenated gases into the subsurface;
4) Vapor extraction to remove vapor-phase contaminants in the subsurface by volatilization of volatile contaminants for subsequent recovery;
5) Construction of hydraulic barriers to mitigate migration of soluble contaminants;
6) Groundwater flushing via injection and recovery of surfactants; and,
7) Pumping/extraction systems for recovery of immiscible and dissolved phase contaminants.

Bioventing

Bioventing is a process of stimulating the natural in situ biodegradation of contaminants in soil by providing air or oxygen to existing soil microorganisms. Bioventing uses low air flow rates to provide only enough oxygen to sustain microbial activity in the unsaturated zone **(Fig 5.2)**. Oxygen is most commonly supplied through direct air injection into residual contamination in soil. In addition to degradation of adsorbed fuel residuals, volatile compounds are biodegraded as vapors move slowly through biologically active soil.

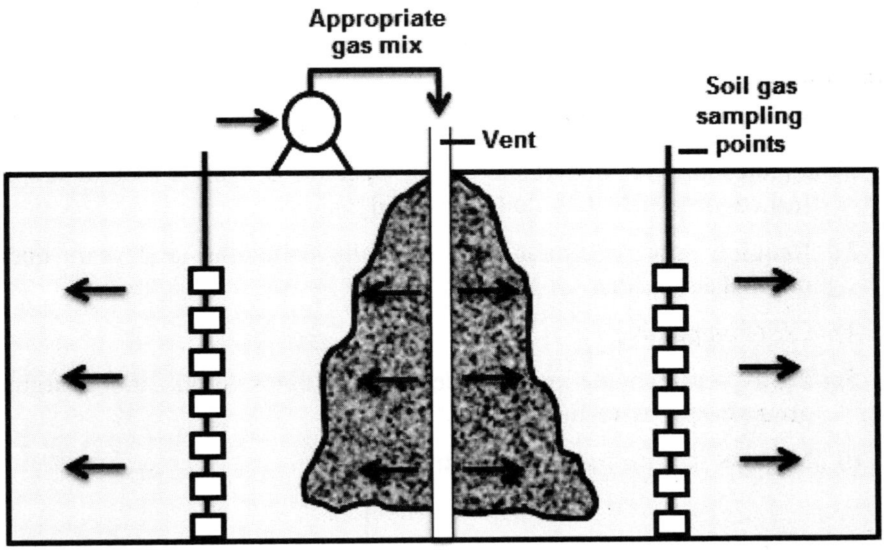

Fig. 5.2: Bioventing

When extraction wells are used for bioventing, the process is similar to soil vapor extraction (SVE). However, while SVE removes constituents primarily through volatilization, bioventing systems promote biodegradation of constituents and minimize volatilization. In practice, some degree of volatilization and biodegradation occurs when either SVE or bioventing is used. Bioventing is applicable to any chemical that can be aerobically biodegraded. Techniques have been successfully used to remediate soils contaminated by petroleum hydrocarbons, non-chlorinated solvents, some pesticides, wood preservatives, and other organic chemicals.

Bioventing is not appropriate for sites with groundwater tables located less than 3 feet below the land surface. Special considerations must be taken for sites with a groundwater table located less than 10 feet below the land surface because groundwater upwelling can occur within bioventing wells under vacuum pressures, potentially occluding screens and reducing or eliminating vacuum-induced soil vapor flow. This potential problem is not encountered if injection wells are used instead of extraction wells to induce air flow. The most important factors that control the effectiveness of bioventing are:

1) The permeability of the petroleum-contaminated soils. This will determine the rate at which oxygen can be supplied to the hydrocarbon-degrading microorganisms found in the subsurface.
2) The biodegradability of the petroleum constituents. This will determine both the rate at which and the degree to which the constituents will be metabolized by microorganisms.

Advantages

1) Uses readily available equipment; easy to install.
2) Creates minimal disturbance to site operations. Can be used to address inaccessible areas (e.g., under buildings).
3) Requires short treatment times: usually 6 months to 2 years under optimal conditions.
4) Cost competitive.
5) Easily combinable with other technologies (e.g., air sparging, groundwater extraction).
6) May not require costly off gas treatment.

Disadvantages

1) High constituent concentrations may initially be toxic to microorganisms.
2) Not applicable for certain site conditions (e.g., low soil permeabilities, high clay content, insufficient delineation of subsurface conditions).
3) Cannot always achieve very low cleanup standards.
4) Permits generally required for nutrient injection wells.
5) Only treats unsaturated-zone soils; other methods may also be needed to treat saturated-zone soils and groundwater.

Cometabolic Bioventing

Cometabolic bioventing has been used at a few sites to treat chlorinated solvents such as TCE, trichloroethane, and dichloroethene (DCE). The equipment used in cometabolic bioventing is similar to aerobic bioventing, but cometabolic bioventing exploits a different biological mechanism. Similar to bioventing, cometabolic bioventing involves the injection of gases into the subsurface; however, cometabolic bioventing injects both air and a volatile organic substrate, such as propane. The concentrations in this gas mixture should be well below the lower explosive limit (LEL), and should be monitored in soil gas. Cometabolic bioventing exploits competitive reactions mediated by monooxygenase enzymes (EPA, 2000). Monooxygenases catalyze the oxidation of hydrocarbons, often through epoxide intermediates, but these enzymes can also catalyze the dechlorination of chlorinated hydrocarbons. Thus, by supplying an appropriate organic substrate and air, cometabolic bioventing can elicit the production of monooxygenases, which consume the organic substrate and facilitate contaminant degradation.

Establishing cometabolic bioventing as the primary mechanism of removal in the field is challenging. Unlike aerobic bioventing, the oxygen use and chlorinated solvent removal are not related stoichiometrically because the metabolism of added organic substrates also consumes oxygen. As a result, measurements of oxygen use, carbon dioxide generation, and contaminant removal cannot be linked stoichiometrically. Indirect measures, such as measuring chloride ion accumulation in the soil and correlating that accumulation to contaminant removal, have been useful at some sites. In addition, collecting data to demonstrate degradation of the organic substrate (by a shutdown test) in the field may be helpful, especially in conjunction with laboratory testing using contaminated soil from the site.

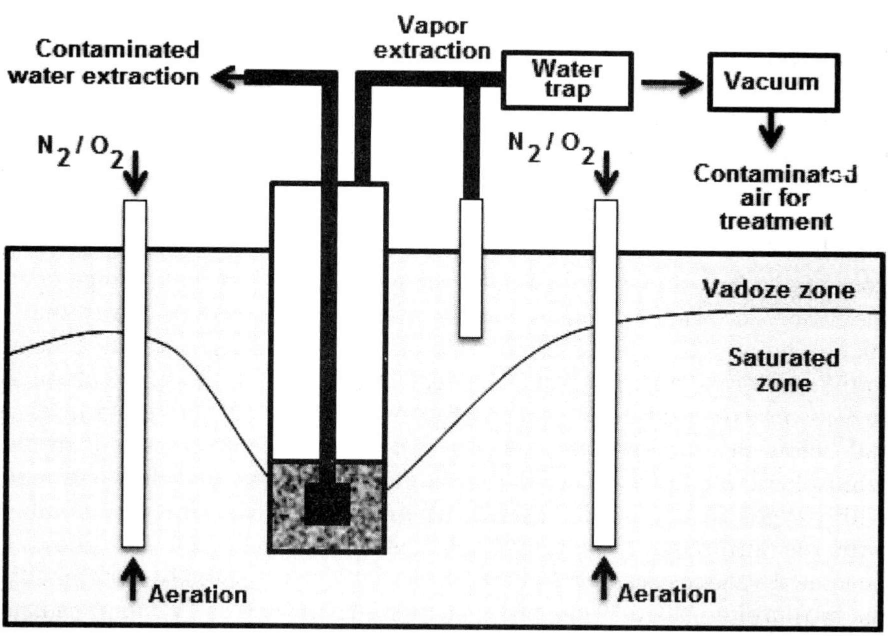

Fig. 5.3: Biosparging

Biosparging

Biosparging is an in situ remediation technology that uses indigenous microorganisms to biodegrade organic constituents in the saturated zone. In Biosparging, air (or oxygen) and nutrients (if needed) are injected into the saturated zone to increase the biological activity of the indigenous microorganisms. Biosparging can be used to reduce concentrations of petroleum constituents that are dissolved in groundwater, adsorbed to soil below the water table, and within the capillary fringe **(Fig 5.3)**.

Although Biosparging can also treat constituents adsorbed to soils in the unsaturated zone, Bioventing is typically more effective for this situation if the indigenous microorganisms are of a high enough concentration to reduce concentrations of petroleum constituents in a timely manner. When used appropriately, Biosparging is effective in reducing petroleum products at underground storage tank (UST) sites. Biosparging is most often used at sites with mid-weight petroleum products (e.g., diesel fuel, jet fuel); lighter petroleum products (e.g., gasoline) tend to volatilize readily and to be removed more rapidly using air Sparging. Heavier products (e.g., lubricating oils) generally take longer to biodegrade than the lighter products, but Biosparging can still be used at these sites. Additionally, one must recognize that it is critical to the process that the exact environment be maintained to insure that the indigenous microorganisms are promoted to grow to biodegrade the contaminants.

Bioslurping

Bioslurping is the only technique serving, on the one hand, the removal of a free contaminant phase swimming on groundwater and, on the other hand, the support of microbial degradation processes. The phase is removed by sucking it off with the aid of a generated vacuum. The bioslurping gauges which are pressure-tight towards atmosphere are equipped with filters in a zone above the groundwater. Within the gauge there is a suction pipe open at the base. The suction pipe is mostly manually adjusted in a way that the suction point will be in the lower zone of the phase. By generating a vacuum in the suction pipe with the aid of a vacuum pump, first of all, the phase and, to a lower extent, also the groundwater is sucked off. If the liquid level will fall below the suction point soil air will be sucked off. This will result in atmospheric air flowing into the unsaturated soil body and thus in a supply with oxygen in conformity with bioventing. Bioslurping may be combined with the infiltration of nutrient salts into the unsaturated soil zone. The generated vacuum ensures that the phase will flow preferably horizontally to the bioslurping gauge. Thus, phase suction (designated as "vacuum enhanced recovery"VER) and bioventing will alternate. As the groundwater level will remain nearly unaffected only little groundwater will be sucked off during bioslurping. Above ground, first of all, the contaminant phase and groundwater will be separated in a liquid separator from the soil air sucked off. The contaminated air is mostly treated by biofiltration and/or activated carbon sorption. The liquid mixture will be split up in an oil/water separator. The contaminant phase obtained may be separately disposed. A treatment of water may be effected e.g. by means of sorption of the contaminants to wet activated carbon **(Fig 5.4)**.

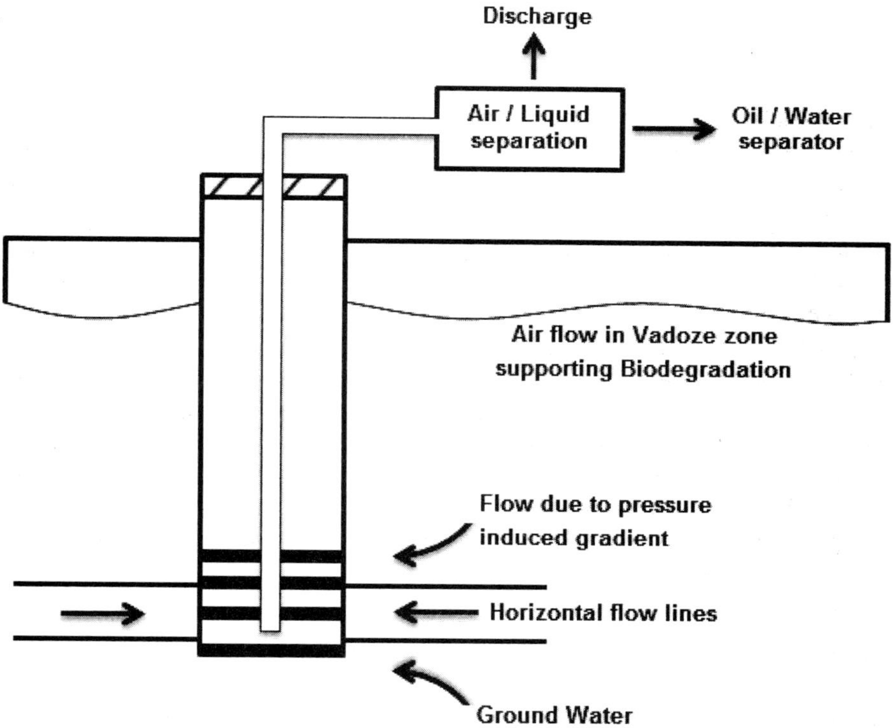

Fig. 5.4: Bioslurping

Applications

1) Especially suited for removing the petroleum product phases.
2) Fine sands up to gravel may be treated.

Advantages

1) Insignificant volume flow of water and air and therefore comparatively small treatment plants and low treatment costs.
2) The main advantage is the horizontally induced flow direction of the phase avoiding a transport of the phase into greater depths.

Disadvantages

1) In particular, if the mixture sucked off contains diesel oil the vacuum pumps may cause water oil emulsions to be formed which may no longer be separated in an oil/water separator.

2) Contaminant phases showing a high viscosity owing to their weathering are only difficult to be treated.

EX SITU BIOREMEDIATION

The conditions necessary for bioremediation in soil cannot always be achieved *in situ*, however. At some sites, the climate may be too cold for microbes to be active, or the soil might be too dense to allow amendments to spread evenly underground. *Ex situ* bioremediation involves removal of waste materials and their collection at a place to facilitate microbial degradation. It suffers from costs associated with solid handling process e.g. excavation, screening and fractionation, mixing, homogenizing and final disposal. *Ex situ* techniques can be faster, easier to control, and used to treat a wider range of contaminants and soil types than in situ techniques. However, they require excavation and treatment of the contaminated soil before and, sometimes, after the actual bioremediation step. On the basis of phases of contaminated materials under treatment *ex situ* bioremediation is classified into two: (i) solid-phase system (including land treatment and soil piles) i.e. composting, and (ii) slurry-phase systems (involving treatment of solid-liquid suspensions in bioreactors).

Solid-phase Bioremediation

Solid-phase bioremediation is a process that treats soils in above-ground treatment areas equipped with collection systems to prevent any contaminant from escaping the treatment. Moisture, heat, nutrients, or oxygen are controlled to enhance biodegradation for the application of this treatment. Solid-phase systems are relatively simple to operate and maintain, require a large amount of space, and cleanups require more time to complete than with slurry-phase processes. Solid-phase soil treatment processes include landfarming, soil biopiles, and composting.

Landfarming

Landfarming, also known as land treatment or land application, is an above-ground remediation technology for soils. It reduces concentrations of contaminants through biodegradation. In the ex-situ process, the contaminated soil is first excavated, mixed with soil amendments such as soil bulking agents and nutrients and then tilled into the earth. The soil is spread over an area and periodically turned to improve aeration. Turning the soil also avoids the disadvantages of having heterogeneous degradation. Soil conditions are controlled to optimize the rate of contaminant degradation. The enhanced microbial activity results in degradation of adsorbed petroleum product constituents through microbial respiration. The petroleum industry has used landfarming for many years. This technique is also applicable in in-situ interventions with a different technological setup.

Landfarming has been proven most successful in treating petroleum hydrocarbons and other less volatile biodegradable contaminants. It can also be applied to certain halogenated volatile, semi-volatile and non-halogenated semi-volatile organic compounds and to pesticides. Diesel fuel, oily sludge, wood-preserving wastes have also been successfully treated. If contaminated soils are shallow (i.e., less than 3 feet below ground surface), it may be possible to effectively stimulate microbial activity without excavating the soils. If petroleum-contaminated soil is deeper than 5 feet, the soils should be excavated and reapplied on the ground surface.

Petroleum products generally contain more than one hundred different constituents that possess a wide range of volatility. In general, gasoline, kerosene, and diesel fuels contain constituents with sufficient volatility to evaporate from a landfarm. Lighter (more volatile) petroleum products (e.g., gasoline) tend to be removed by evaporation during landfarm aeration processes (i.e., tilling or plowing) and, to a lesser extent, degraded by microbial respiration. Depending upon regulations for air emissions of volatile organic compounds (VOCs), emissions of VOCs should be controlled. Control involves capturing the vapors before they are emitted to the atmosphere, passing them through an appropriate treatment process, and then venting them to the atmosphere.

Process involved in landfarming

Soil normally contains large numbers of diverse microorganisms including bacteria, algae, fungi, protozoa, and actinomycetes. In well-drained soils, which are most appropriate for landfarming, these organisms are generally aerobic. Of these organisms, bacteria are the most numerous and biochemically active group, particularly at low oxygen levels. Bacteria require a carbon source for cell growth and an energy source to sustain metabolic functions required for growth. Bacteria also require nitrogen and phosphorus for cell growth. Although sufficient types and quantities of microorganisms are usually present in the soil, recent applications of ex-situ soil treatment include blending the soil with cultured microorganisms or animal manure. Incorporating manure serves to both augment the microbial population and provide additional nutrients.

The effectiveness of landfarming depends on parameters that may be grouped into three categories:

1) Soil characteristics

2) Constituent characteristics

3) Climatic conditions.

Soil texture affects the permeability, moisture content, and bulk density of the soil. To ensure that oxygen addition (by tilling or plowing), nutrient distribution, and moisture content of the soils can be maintained within effective ranges, texture of the soils should be analyzed. For example, soils which tend to clump together (such as clays) are difficult to aerate and result in low oxygen concentrations. It is also difficult to uniformly distribute nutrients throughout these soils. They also retain water for extended periods following a precipitation event.

Typical landfarms are uncovered and, therefore, exposed to climatic factors including rainfall, snow, and wind, as well as ambient temperatures. Rainwater that falls directly onto, or runs onto, the landfarm area will increase the moisture content of the soil and cause erosion. During and following a significant precipitation event, the moisture content of the soils may be temporarily in excess of that required for effective bacterial activity. On the other hand, during periods of drought, moisture content may be below the effective range and additional moisture may need to be added. Erosion of landfarm soils can occur during windy periods and particularly during tilling or plowing operations. Wind erosion can be limited by plowing soils into windrows and applying moisture periodically.

Landfarm Construction

Landfarm Construction includes steps like site preparation (grubbing, clearing and grading); berms; liners (if necessary); leachate collection and treatment systems; soil pretreatment methods (e.g., shredding, blending and amendments for fluffing, pH control); and enclosures and appropriate vapor treatment facilities **(Fig 5.5)**.

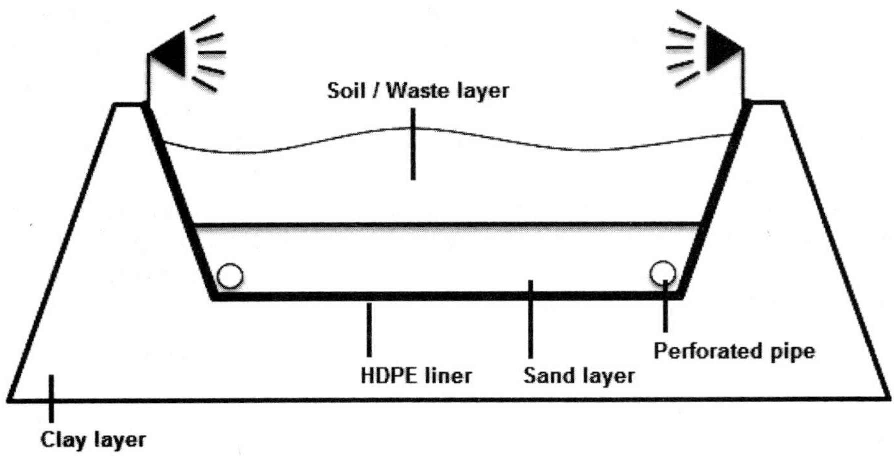

Fig. 5.5: Landfarming for treatment of contaminated waste or soil

If the site is located in an area subject to annual rainfall of greater than 30 inches during the landfarming season, a rain shield (such as a tarp, plastic tunnel, or greenhouse structure) should be considered in the design of the landfarm. In addition, rainfall run-on and run-off from the landfarm should be controlled using berms at the perimeter of the landfarm. A leachate collection system at the bottom of the landfarm and a leachate treatment system may also be necessary to prevent groundwater contamination from the landfarm.

Microorganisms require inorganic nutrients such as nitrogen and phosphorus to support cell growth and sustain biodegradation processes. Nutrients may be available in sufficient quantities in the site soils but, more frequently, nutrients need to be added to landfarm soils to maintain bacterial populations. However, excessive amounts of certain nutrients (i.e., phosphate and sulfate) can repress microbial metabolism.

It is important to make sure that system operation and monitoring plans have been developed for the landfarming operation. Regular monitoring is necessary to ensure optimization of biodegradation rates, to track constituent concentration reductions, and to monitor vapor emissions, migration of constituents into soils beneath the landfarm (if unlined), and groundwater quality.

Advantages

1) It is extremely simple and rather inexpensive and requires no process controls.
2) Relatively unskilled personnel can perform the technique.
3) Certain pollutants can be completely removed from the soil.

Disadvantages

1) It requires extensive space and time.
2) Certain pollutants cannot be reduced to sufficiently low levels.
3) Runoff must be collected and may require treatment.
4) It can incorporate contaminated soil into uncontaminated soil, creating a larger volume of contaminated material.
5) Conditions affecting biological degradation of contaminants (e.g. temperature and rainfall) are largely uncontrolled. This may increase the time to complete remediation.
6) Presence of significant heavy metal concentrations (greater than 2,500 ppm) may inhibit microbial growth.
7) Volatile constituents tend to evaporate rather than biodegrade during treatment.

Biopiles

Biopiles, also known as biocells, bioheaps, biomounds, and compost piles, are used to reduce concentrations of petroleum constituents in excavated soils through the use of biodegradation. This technology involves heaping contaminated soils into piles (or "cells") and stimulating aerobic microbial activity within the soils through the aeration and/or addition of minerals, nutrients, and moisture **(Fig 5.6)**. The enhanced microbial activity results in degradation of adsorbed petroleum-product constituents through microbial respiration. Biopiles are similar to landfarms in that they are both above-ground, engineered systems that use oxygen, generally from air, to stimulate the growth and reproduction of aerobic bacteria which, in turn, degrade the petroleum constituents adsorbed to soil. While landfarms are aerated by tilling or plowing, biopiles are aerated most often by forcing air to move by injection or extraction through slotted or perforated piping placed throughout the pile.

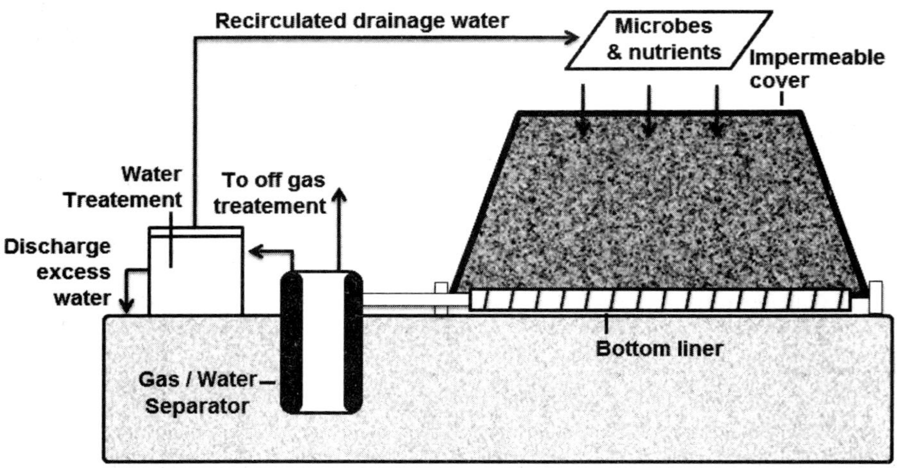

Fig. 5.6: Biopiles for waste treatment

Biopiles, like landfarms, have been proven effective in reducing concentrations of nearly all the constituents of petroleum products typically found at underground storage tank (UST) sites. Lighter (more volatile) petroleum products (e.g., gasoline) tend to be removed by evaporation during aeration processes (i.e., air injection, air extraction, or pile turning) and, to a lesser extent, degraded by microbial respiration.

Biopiles are designed to optimize the conditions for aerobic bacteria to biodegrade organic contaminants. The effectiveness of a biopile system depends on many parameters which can be grouped into three categories:

1) Soil characteristics.

2) Constituent characteristics.

3) Climatic conditions.

Advantages

1) Very simple technology to design and implement
2) Cost competitive technology
3) Can be designed to be a closed system
4) Can be engineered to be potentially effective for any combination of site conditions and petroleum products.

Disadvantages

1. Presence of significant heavy metal concentrations may inhibit microbial growth.
2. Vapor generation during aeration may require treatment prior to discharge.
3. Contaminated soils must be excavated and dust and noise must be controlled
4. Static treatment processes may result in less uniform treatment than processes that involve periodic mixing.

Composting

Composting is a controlled biological process by which organic contaminants in the soil are converted by microorganisms, under both aerobic and anaerobic conditions, to innocuous, stabilized byproducts. Soils are excavated and mixed with bulking agents and organic amendments such as wood chips and plant wastes. Proper conditions of oxygen and moisture help to achieve maximum degradation efficiency **(Fig 5.7)**.

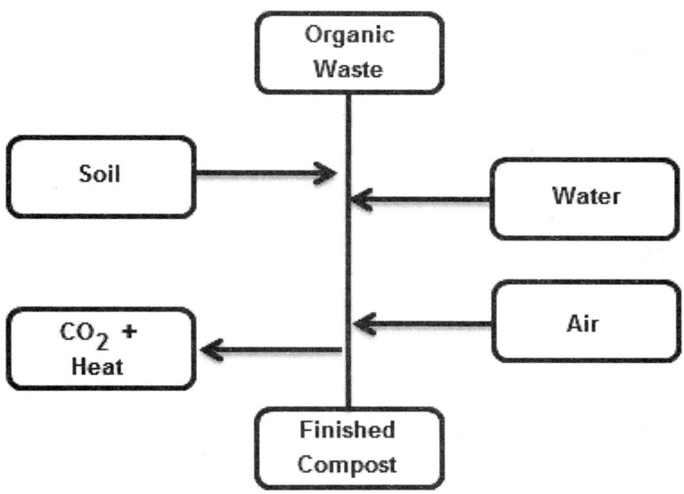

Fig. 5.7: General steps of composting

However, many hazardous compounds are resistant to microbial degradation due to complex chemical structure, toxicity and compound concentration that hardly support growth. Microbial growth is also affected by moisture, pH, inorganic nutrients and particle size. Because composting of hazardous wastes typically involves the bioremediation of contaminated substrate-sparse soils, support of microbial self-heating needs incorporation of proper amount of supplements. The hazardous compounds reported to disappear through composting includes aliphatic and aromatic hydrocarbons and certain halogenated compounds. The possible routes leading to disappearance of hazardous compounds include volatilization, assimilation, adsorption, polymerization and leaching.

There are three major designs used in composting:

1) Aerobic static pile – compost is formed into piles and aerated with blowers or vacuum pumps.
2) Mechanically agitated in-vessel composting- compost is placed in a reactor vessel where it is mixed and aerated.
3) Windrow composting- compost is placed in long piles known as windrows and periodically mixed with mobile equipment.

Slurry-phase Bioremediation

Bioreactors are favored over in situ biological techniques for heterogeneous soils, low permeability soils, areas where underlying ground water would be difficult to capture, or when faster treatment times are required. Slurry-phase bioreactors are used primarily to treat non-halogenated SVOCs and VOCs in excavated soils or dredged sediments.

Slurry phase biological treatment involves the controlled treatment of excavated soil in a bioreactor. The excavated soil is first processed to physically separate stones and rubble. The soil is then mixed with water to a predetermined concentration dependent upon the concentration of the contaminants, the rate of biodegradation, and the physical nature of the soils. Some processes pre-wash the soil to concentrate the contaminants. Clean sand may then be discharged, leaving only contaminated fines and wash water to biotreat. Typically, slurry contains from 10 to 30% solids by weight.

The solids are maintained in suspension in a reactor vessel and mixed with nutrients and oxygen. If necessary, an acid or alkali may be added to control pH. Microorganisms also may be added if a suitable population is not present. When biodegradation is complete, the soil slurry is dewatered. Dewatering devices that may be used include clarifiers, pressure filters, vacuum filters, sand drying beds, or centrifuges.

Slurry-phase biological treatment can be a relatively rapid process compared to other biological treatment processes, particularly for contaminated clays. The success of the process is highly dependent on the specific soil and chemical properties of the contaminated material. This technology is particularly useful where rapid remediation is a high priority. Biologically there are three types of slurry-phase bioreactors: aerated lagoons, low-shear airlift reactor, and fluidized-bed soil reactor.

Limitations

a) Factors that may limit the applicability and effectiveness of the slurry-phase biotreatment process include:

b) Excavation of contaminated media is required, except for lagoon implementation.

c) Sizing of materials prior to putting them into the reactor can be difficult and expensive. Non-homogeneous soils and clayey soils can create serious materials handling problems. In the case of free phase contaminant, precluded removal is mandatory.

d) Dewatering soil fines after treatment can be expensive.

e) An acceptable method for disposing of non-recycled wastewaters is required.

Bioaugmentation

Bioaugmentation is the application of specifically selected bacteria (microbes) into a wastewater treatment system to enhance the system performance in some way. Bioaugmentation has many applications in all types of biological wastewater treatment systems, including once-through lagoons, activated sludge plants, sequencing batch reactors (SBR's) and rotating biological contactors (RBC's). Because microorganisms are the heart of any biological wastewater system, it makes sense that by enhancing the microbial community, the overall wastewater system can operate more efficiently. The purpose of bioaugmentation is to supplement the existing microbial community in order to improve its functionality. Bioaugmentation offers many advantages over traditional technology platforms like chemicals, equipments, or other consumables, and has been used in secondary wastewater treatment systems for decades.

Bioremediation of hydrocarbons is best accomplished with a process called Bio-augmentation. That is, the addition of a large number of selected, laboratory grown standardized microorganisms to a contaminated matrix. These contaminant specific microbes capable of degrading these hydrocarbon compounds breaking them down into carbon dioxide and water. The microbes will survive and consume their contaminant food source until the unwanted pollutant is remediated. This is possible because the microbes use parts of the hydrocarbons to maintain their own metabolic process.

Bioaugmentation can be used for in-situ and ex-situ soil and water remediation. It can be used in conjunction with traditional mechanical methods or on its own. Microbial bioaugmentation is most successful in treating the residual biofilms in the soil.

BIOREMEDIATION OF PETROLEUM HYDROCARBONS

Petroleum compounds consist of four fractions: saturated hydrocarbons, aromatic hydrocarbons, nitrogen-sulphur-oxygen containing compounds and asphaltenes. Normally, of the saturated hydrocarbons, the straight-chain n-alkanes are most susceptible to biodegradation, whereas branched alkanes are less vulnerable to microbial attack. The aromatic fraction is more difficult to biodegrade and the susceptibility of its components decreases as the number of aromatic or alicyclic rings in the molecule increases. Polycyclic (polynuclear) aromatic hydrocarbons occur extensively as pollutants in soil and water and are important environmental contaminants because of their recalcitrance. These compounds also constitute a potential risk to human health, as many of them are carcinogens. Microorganisms that biodegrade the components of petroleum hydrocarbons are isolated from various environments, particularly from petroleum-contaminated sites. Evaluations of indigenous microorganisms are needed so that bacterial community composition can be correlated with ability to degrade target pollutants. Certain plasmids play an important role in adaptation of natural microbial populations to oil and other hydrocarbons.

Bacteria usually are the dominant hydrocarbon degraders in aquatic systems such as oceans. They also possess diverse metabolic pathways that are not present in fungi which allow them to utilize most recalcitrant petroleum hydrocarbons. Bacterial degradation of aromatic compounds can be divided into three steps:

(i) Modification and conversion of the many different compounds into a few central aromatic intermediates (ring-fission substrates); this step is referred as peripheral pathway and involves considerable modification of the ring and/or perhaps elimination of substituent groups;

(ii) Oxidative ring cleavage by dioxygenases, which are responsible for the oxygenolytic ring cleavage of dihdyroxylated aromatic compounds (catechol, protocatechuate, gentisate);

(iii) Further degradation of the non-cyclic, non-aromatic ring-fission products to intermediates of central metabolic pathways.

Long-chain hydrocarbons (C_{10}-C_{18}) can be used rapidly by many high G+C Gram-positive bacteria. Only a few bacteria can oxidize C_2-C_8 hydrocarbons. Degradation of n-alkanes requires activation of the inert substrates by molecular oxygen with help of oxygenases by three possible ways that are associated with membranes **(Fig 5.8)**:

Mono oxygenase attacks at the end producing alkan-1-ol:

$$\mathbf{R\text{-}CH_3 + O_2 + NAD(P)H + H^+ \rightarrow R\text{-}CH_2OH + NAD(P)^+ + H_2}$$

Dioxygenase attack produces the hydro peroxides, which are reduced to yield also alkan-1-ol:

$$\mathbf{R\text{-}CH_3 + O_2 \rightarrow R\text{-}CH_2\,COOH + NAD(P)H + H^+ \rightarrow R\text{-}CH_2OH + NAD(P)^+ + H_2O}$$

Rarely, subterminal oxidation at C_2 by mono oxygenase yields secondary alcohols.

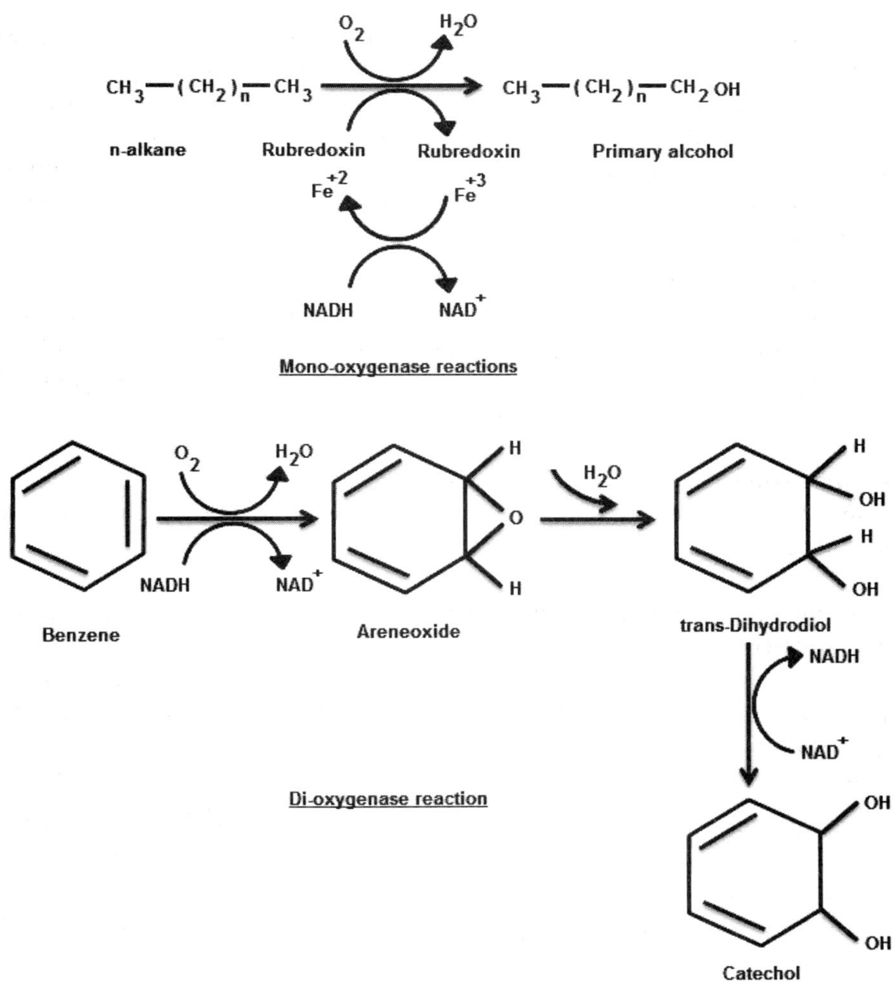

Fig. 5.8: Degradation of Hydrocarbons

Genetically Engineered Microbes

Various genes responsible for degradation of environmental pollutants, for example, toluene, chlorobenzene acids, and other halogenated pesticides and toxic wastes have been identified. For every compound, one separate plasmid is required. It is not like that one plasmid can degrade all the toxic compounds of different groups. The plasmids are grouped into four categories:

a) OCT plasmid which degrades, octane, hexane and decane.

b) XYL plasmid which degrades xylene and toluenes.

c) CAM plasmid that decompose camphor.

d) NAH plasmid which degrades naphthalene.

Dr Anand Mohan Chakrabarty (an Indian born American scientist) produced a new product of genetic engineering called as superbug (oil eating bug) by introducing plasmids from different strains into a single cell of *P. putida*. This superbug is such that can degrade all the four types of substrates for which four separate plasmids were required. There are various microorganisms used for bioremediation of petroleum based hydrocarbons. Some of them are as follows:

Pseudomonas putida

Pseudomonas putida is a gram-negative soil bacterium that is involved in the bioremediation of toulene, a component of paint thinner. It is also capable of degrading naphthalene, a product of petroleum refining, in contaminated soils.

Dechloromonas aromatica

A soil bacteria genus which are capable of degrading perchlorate and aromatic compounds.

Nitrosomonas europaea*, *Nitrobacter hamburgensis*, and *Paracoccus denitrificans

The removal of nitrogen is a two stage process that involves nitrification and denitrification. During nitrification, ammonium is oxidized to nitrite by organisms like *Nitrosomonas europaea*. Then, nitrite is further oxidized by microbes like *Nitrobacter hamburgensis*.

In anaerobic conditions, nitrate produced during ammonium oxidation is used as a terminal electron acceptor by microbes like *Paracoccus denitrificans*. The result is dinitrogen gas. Through this process, ammonium and nitrate, two pollutants responsible for eutrophication in natural waters, are remediated.

Phanerochaete chrysosporium

The lignin-degrading white rot fungus, *Phanerochaete chrysosporium*, exhibits strong potential for bioremediation of: pesticides, polyaromatic hydrocarbons, PCBs, dioxins, dyes, TNT and other nitro explosives, cyanides, azide, carbon tetrachloride, and pentachlorophenol. White rot fungi degrade lignin with nonselective extracellular peroxidases, which can also facilitate the degradation of other compounds containing similar structure to lignin within the proximity of the enzymes released.

Deinococcus radiodurans

Deinococcus radiodurans is a radiation-resistant extremophile bacterium that is genetically engineered for the bioremediation of solvents and heavy metals. An engineered stain of *Deinococcus radiodurans* has been shown to degrade ionic mercury and toluene in radioactive mixed waste environments.

Methylibium petroleiphilum

Methylibium petroleiphilum (formally known as PM1 strain) is a bacterium is capable of methyl tert-butyl ether (MTBE) bioremediation. PM1 degrades MTBE by using the contaminant as the sole carbon and energy source.

PHYTOREMEDIATION

Phytoremediation consists of mitigating pollutant concentrations in contaminated soils, water, or air, with plants able to contain, degrade, or eliminate metals, pesticides, solvents, explosives, crude oil and its derivatives, and various other contaminants from the media that contain them. Phytoremediation may be applied wherever the soil or static water environment has become polluted or is suffering ongoing chronic pollution. Examples where phytoremediation has been used successfully include the restoration of abandoned metal-mine workings, reducing the impact of sites where polychlorinated biphenyls have been dumped during manufacture and mitigation of on-going coal mine discharges.

Phytoremediation refers to the natural ability of certain plants called hyperaccumulators to bioaccumulate, degrade,or render harmless contaminants in soils, water, or air. Contaminants such as metals, pesticides, solvents, explosives, and crude oil and its derivatives, have been mitigated in phytoremediation projects worldwide. Many plants such as mustard plants, alpine pennycress, hemp, and pigweed have proven to be successful at hyperaccumulating contaminants at toxic waste sites.

Types of Phytoremediation

Phytoremediation can be classified in to subcategories depending up on the type of remediation **(Fig 5.9)**:

Rhizodegradation

Rhizodegradation is the enhancement of naturally-occurring biodegradation in soil through the influence of plant roots, and ideally will lead to destruction or detoxification of an organic contaminant. A wide range of organic contaminants are candidates for rhizodegradation, such as petroleum hydrocarbons, PAHs, pesticides, polychlorinated biphenyls (PCBs), surfactants and chlorinated solvents.

Phytodegradation

Phytodegradation, also called as phytotransformation, is the uptake, metabolizing, and degradation of contaminants within the plant, or the degradation of contaminants in the soil, sediments, sludges, ground water, or surface water by enzymes produced and released by the plant. Chlorinated solvents like TCE, some organic herbicides and trinitrotoluene can be degraded using this method.

Phytoextraction

Phytoextraction (also known as phytoaccumulation, phytoabsorption, and phytosequestration) is contaminant uptake by roots with subsequent accumulation in the above ground portion of a plant, generally to be followed by harvest and ultimate disposal of the plant biomass. Phytoextraction has also been referred to as phytomining or biomining. Phytomining is the use of plants to obtain a gain from hyperaccumulated metals extracted by a plant, whether from contaminated soils or from soils having naturally high concentrations of metals. This is particularly useful for removing metals from soil and, in some cases; incorporation of plant incinerations will help metal reuse. These processes extract both metallic and organic constituents from soil by direct uptake into plants and translocation to aboveground biomass using metal- (hyper) accumulating plants. *Brassica juncea, Berkeya coddii, Allysum bertolonii, Thlaspi caerulescens* and *Thlaspi goesingense* are some of the plants involved in phytoextraction. The main advantage of phytoextraction is the process is eco- friendly but will take more time than anthropogenic soil clean-up methods.

Rhizofiltration

Rhizofiltration (also known as phytofiltration) is the removal by plant roots of contaminants in surface water, waste water, or extracted ground water, through adsorption or precipitation onto the roots, or absorption into the roots. Here accumulation can occur in root or can be retained in any portion of the plant. Plants used for rhizofiltration are not planted directly *in situ* but are acclimated to the pollutant first, which makes the process little tedious and time consuming. Sunflowers grown in radioactively contaminated pools exemplify this process.

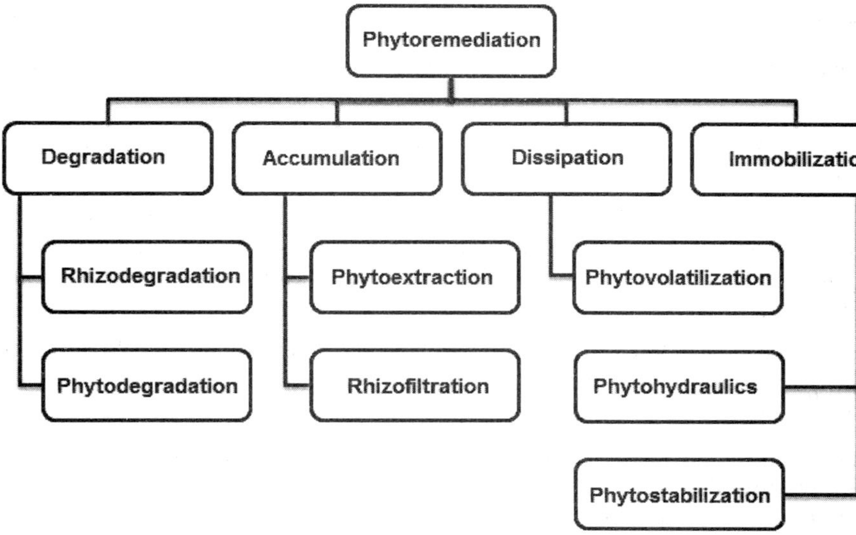

Fig. 5.9: Types of Phytoremediation

Phytovolatilization

Phytovolatilization is the uptake of a water soluble contaminant by a plant, and the subsequent release of a volatile contaminant, a volatile degradation product of a contaminant, or a volatile form of an initially non-volatile contaminant.

Phytostabilization

Phytostabilization (also called as phytoimmobilization) is the use of plants to immobilize soil and water contaminants. Some organic contaminants or metabolic byproducts of these contaminants can be attached or incorporated into plant components such as lignin and such type of phytostabilization is called phytolignification. Indian mustard appeared to have potential for phytostabilization.

Phytohydraulics

Phytohydraulics is theuse of deep-rooted plants to degrade ground water contaminants that come into contact with their roots. Ground water plume of methyl-tert-butyl-ether (MTBE) has been recovered using this technique.

REFERENCES

Alexander, M., Biodegradation and Bioremediation. San Diego, CA:Academic Press, 1994.

Ada, OK: U.S. Environmental Protection Agency, Office of Research and Development. EPA/5R-93/124, 1993.

Alexander, S.K. and Webb, J.W. (1987) Relationship of Spartina alterniflora growth to sediment oil content following an oil spill. Proceedings of 1987 International Oil Spill Conference. American Petroleum Institute, Washington, DC, pp 445-449.

Alexander, S.K. and Webb, J.W. (1985) Seasonal response of Spartina alterniflora to oil. Proceedings of 1985 International Oil Spill Conference. American Petroleum Institute, Washington, DC, pp 355-357.

Alexander, S.K. and Webb, J.W. (1983) Effects of oil on growth and decomposition of Spartina alterniflora. Proceedings of 1983 International Oil Spill Conference. American Petroleum Institute, Washington, DC.

ATI Orion (1991a) Instruction manual platinum redox electrodes model 96-78-00 model 97-78-00. Orion Research Inc., Boston.

ATI Orion (1991b) Model 94-16 silver/sulfide electrode instruction manual. Orion Research Inc., Boston.

Atlas, R.M. and Bartha R. (1992) Hydrocarbon biodegradation and oil spill bioremediation. In K.C. Marshall (ed.), Advances in Microbial Ecology, Vol. 12, Plenum Press, NY, pp287-338.

AZUR Environmental. (1999) MicrotoxOmni for Windows 95/98, ver. 1.18, CD-ROM. AZUR Environmental, Carlsbad CA.

Bachoon, D.S., Hodson, R.E., Araujo, R. (2001) Microbial community assessment in oil-impacted salt marsh sediment microcosms by traditional and nucleic acid-based indices. Journal of Microbial Methods, 46, 37-49.

Baker, J.M. (1999) Ecological effectiveness of oil spill countermeasures: how clean is clean? Pure Appl. Chem., 71(1), 135-151.

Baker, J. M., Guzman, L. M., Bartlett, P. D., Little, D. I., and Wilson, C. M. (1993). Long-term fate and effects of untreated thick oil deposits on salt marshes. Proceedings of 1993 International Oil Spill Conference, American Petroleum Institute, Washington, D.C., pp. 395-399.

Baker, J,M, Bayley, J.A., Howells, S.E., Oldham, J., Wilson, M. (1989) Oil in wetland. In B. Dicks (ed): Ecological Impacts of the Oil Industry, John Wiley & Sons Ltd, Chichester, U.K., pp 37-59.

Banks, M.K., Govindaraju, R. S., Schwab, A. P., Kulakow, P., and Finn, J. (2000). Phytoremediation of Hydrocarbon-Contaminated Soil. Lewis Publishers, Boca Rotan, FL.

Bergen, A., Alderson C., Bergfors, R., Aquila, C., Matsil, M.A. (2000) Restoration of a spartina alterniflora salt marsh following a fuel oil spill, New York City, NY. Wetlands Ecology and Management, 8, 185-195.

Coates, J.D., Woodward, J., Allen, J., Philip, P., Lovley, D.R. (1997) Anaerobic degradation of polycyclic hydrocarbons and alkanes in petroleum-contaminated marine harbour sediments. Appl. Environ. Microbiol., 63, 3589-3593.

Cunningham, S.D., Anderson, T.A., Schwab, A.P., and Hsu, F.C. (1996) Phytoremediation of soils contaminated with organic pollutants, Advances in Agronomy, 56, 55-114.

Fan, C.Y. and A.N. Tafuri. "Engineering Application of Biooxidation Processes for Treating Petroleum-Contaminated Soil," in D.L. Wise and D.J. Trantolo, eds. Remediation of Hazardous Waste Contaminated Soils. New York, NY: Marcel Dekker, Inc., pp. 373-401,1994.

Flathman, P.E. and D.E. Jerger. Bioremediation Field Experience. Boca Raton, FL: CRC Press, 1993.

Freeman, H.M. Standard Handbook of Hazardous Waste Treatment and Disposal. New York, NY: McGraw-Hill Book Company, 1989.

Garcia-Blanco, S., Suidan, M.T., Venosa, A.D., Huang, T., Cacho-Rivero, J. (2001a) Microcosm study of effect of different nutrient addition on bioremediation of fuel oil #2 in soil from Nova Scotia coastal marshes. Proceedings of 2001 International Oil Spill Conference. American Petroleum Institute, Washington DC.

Garcia-Blanco, S., Motelab, M., Venosa, A.D., Suidan, M.T., Lee, K., King, D.W. (2001b) Restoration of the oil-contaminated Saint Lawrence River shoreline: Bioremediation and Phytoremediation. Proceedings of 2001 International Oil Spill Conference. American Petroleum Institute, Washington DC, pp. 303-308.

Gauthier J. (2001) Bioremediation of a crude oil-contaminated riverine shoreline : Sediment toxicity assessment with ASPA an algal solid phase (direct contact) assay. Maurice Lamontagne Institute, Fisheries and Oceans Canada. 15p.

Getter C.D. and Ballou, T.G. (1985) Field experiments on the effects of oil and dispersant on mangroves. Proceedings of 1985 International Oil Spill Conference. American Petroleum Institute, Washington DC, pp577-582.

Gilfillan, E. S., Suchanek, T. H., Boehm, P. D., Harner, E. J., Page, D. S., and Sloan, N. A. (1995) Shoreline impacts in the Gulf of Alaska region following the Exxon Valdez oil spill. In: Exxon Valdez Oil Spill: Fate and Effects in Alaskan Waters. P. G. Wells, J. N. Butler and J. S. Hughes (eds.), American Society for Testing and Materials, Philadelphia, ASTM STP 1219. pp. 444-481.

Grasso, D. Hazardous Waste Site Remediation, Source Control. Boca Raton, FL: CRC Press, 1993.

Greene, T.C. (1991) The Apex Barges Spill, Galveston Bay , July 1990. Proceedings of 1991 International Oil Spill Conference. American Petroleum Institute, Washington DC, pp291-297.

Norris, R.D., Hinchee, R.E., Brown, R.A., McCarty, P.L., Semprini, L.,Wilson, J.T., Kampbell, D.H., Reinhard, M., Bower, E.J., Borden, R.C.,Vogel, T.M., Thomas, J.M., and C.H. Ward. Handbook of Bioremediation. Boca Raton, FL:CRC Press, 1994.

Norris, R.D., Hinchee, R.E., Brown, R.A., McCarty, P.L., Semprini, L.,Wilson, J.T., Kampbell, D.H., Reinhard, M., Bower, E.J., Borden, R.C.,Vogel, T.M., Thomas, J.M., and C.H. Ward. In-Situ Bioremediation of Ground Water and Geological Material: A Review of Technologies.

Pope, Daniel F., and J.E. Matthews. Environmental Regulations and Technology: Bioremediation Using the Land Treatment Concept. Ada, OK: U.S. Environmental Protection Agency, Environmental Research Laboratory. EPA/600/R-93/164, 1993.

Wright, A.L., Weaver, R.W., Webb, J.W. (1996) Concentrations of N and P in floodwater and uptake of 15N by spartina alterniflora in oil-contaminated mesocosms. Bioresource Technology, 56, 257-264.

Zambon, S., N.M. Fragoso, S.P. Tabash, S.M. Billiard, and P.V. Hodson. (2000) The bioavailability and toxicity of sediment-borne retene, in Proceedings of the 21st Annual Meeting of the Society Environmental Toxicology and Chemistry, Nashville, TN, Nov. 12-16, 2000, Abstract PTA 083.

Zhu, X., Venosa, A.D., Suidan, M.T., and Lee, K. (2001) Guidelines for the Bioremediation of Marine Shorelines and Freshwater Wetlands, Report under a contract with Office of Research and Development, U.S. Environmental Protection Agency. Available on-line at: http://www.epa.gov/oilspill/pdfs/bioremed.pdf

Zobell, C.E. (1973) Microbial degradation of oil: Present statue, problems, and perspectives. In Ahearn and Meyers (Eds.), The Microbial Degradation of Oil Pollutants, Publication No. LSU-SG-73-01, Louisiana State University, Baton Rouge, LA, pp3-16.

6

Solid Waste Management

Since the beginning, humankind has been generating waste, be it the bones and other parts of animals they slaughter for their food or the wood they cut to make their carts. With the progress of civilization, the waste generated became of a more complex nature. At the end of the 19th century the industrial revolution saw the rise of the world of consumers. Not only did the air get more and more polluted but the earth itself became more polluted with the generation of nonbiodegradable solid waste. The increase in population and urbanization was also largely responsible for the increase in solid waste. Solid waste means any garbage, refuse, sludge from a wastewater treatment plant, water supply treatment plant, or air pollution control facility and other discarded materials including solid, liquid, semi-solid, or contained gaseous material, resulting from industrial, commercial, mining and agricultural operations, and from community activities. Solid wastes are any discarded (abandoned or considered waste-like) materials. Solid wastes can be solid, liquid, and semi-solid or containerized gaseous material.

Each household generates garbage or waste day in and day out. Items that we no longer need or do not have any further use for fall in the category of waste, and we tend to throw them away. There are different types of solid waste depending on their source. In today's polluted world, learning the correct methods of handling the waste generated has become essential. Segregation is an important method of handling municipal solid waste. Segregation at source can be understood clearly by schematic representation. One of the important methods of managing and treating wastes is composting. As the cities are growing in size and in problems such as the generation of plastic waste, various municipal waste treatment and disposal methods are now being used to try and resolve these problems. One common sight in all cities is the rag picker who plays an important role in the segregation of this waste. Garbage generated in households can be recycled and reused to prevent creation of waste at source and reducing amount of waste thrown into the community dustbins. Garbage is what someone leaves behind that they do not want to use anymore. It can also be called waste or rubbish. A definition of garbage is anything left behind at a place where you used to be, but are not anymore.

The broad categories of garbage are as follows:

(i) Organic waste: kitchen waste, vegetables, flowers, leaves, fruits.

(ii) Toxic waste: old medicines, paints, chemicals, bulbs, spray cans, fertilizer and pesticide containers, batteries, shoe polish.

(iii) Recyclable: paper, glass, metals, plastics.

(iv) Soiled: hospital waste such as cloth soiled with blood and other body fluids.

(v) Discarded appliances and vehicles.

(vi) Uncontaminated used oil and anti-freeze empty aerosol cans, paint cans and compressed gas cylinders construction and demolition debris, asbestos etc.

TYPES OF SOLID WASTE

Waste can be classified in several ways but the following list represents a typical classification:

a) Biodegradable waste: food and kitchen waste, green waste, paper (can also be recycled).

b) Recyclable material: paper, glass, bottles, cans, metals, certain plastics, fabrics, clothes, batteries etc.

c) Inert waste: construction and demolition waste, dirt, rocks, debris.

d) Electrical and electronic waste - electrical appliances, TVs, computers, screens, etc.

e) Composite wastes: waste clothing, Tetra Packs, waste plastics such as toys.

f) Hazardous waste including most paints, chemicals, light bulbs, fluorescent tubes, sprays cans, fertilizer and containers.

g) Toxic waste including pesticide, herbicides, fungicides.

h) Medical waste.

Solid waste can also be categories into three main categories depending upon their source of generation.

a) Household waste is generally classified as municipal waste.

b) Industrial waste as hazardous waste.

c) Biomedical waste or hospital waste as infectious waste.

Table 6.1: Sources and types of waste

SOURCES	TYPICAL WASTE GENERATORS	COMPONENTS OF SOLID WASTE
Residential	Single and multifamily dwellings	Food wastes, paper, cardboard, plastics, textiles, glass, metals, ashes, special wastes (bulky items, consumer electronics, batteries, oil, tires) and household hazardous wastes
Commercial	Stores, hotels, restaurants, markets, office buildings	Paper, cardboard, plastics, wood, food wastes, glass, metals, special wastes, hazardous wastes
Institutional	Schools, government center, hospitals, prisons	Paper, cardboard, plastics, wood, food wastes, glass, metals, special wastes, hazardous wastes
Municipal	Street cleaning, landscaping, parks, beaches, recreational areas	Street sweepings, landscape and tree trimmings, general wastes from parks, beaches, and other recreational areas

Municipal Solid Waste

Municipal solid waste consists of household waste, construction and demolition debris, sanitation residue, and waste from streets. This garbage is generated mainly from residential and commercial complexes. With rising urbanization and change in lifestyle and food habits, the amount of municipal solid waste has been increasing rapidly and its composition changing. Over the last few years, the consumer market has grown rapidly leading to products being packed in cans, aluminum foils, plastics, and other such nonbiodegradable items that cause incalculable harm to the environment. In India, some municipal areas have banned the use of plastics and they seem to have achieved success.

Hazardous Waste

Industrial and hospital waste is considered hazardous as they may contain toxic substances. Certain types of household waste are also hazardous. Hazardous wastes could be highly toxic to humans, animals, and plants; are corrosive, highly inflammable, or explosive; and react when exposed to certain things e.g. gases. Household wastes that can be categorized as hazardous waste include old batteries, shoe polish, paint tins, old medicines, and medicine bottles. Hospital waste contaminated by chemicals used in hospitals is considered hazardous. These chemicals include formaldehyde and phenols, which are used as disinfectants, and mercury, which is used in thermometers or equipment that measure blood pressure. In the industrial sector, the major generators of hazardous waste are the metal, chemical, paper, pesticide, dye, refining, and rubber goods industries. Direct exposure to chemicals in hazardous waste such as mercury and cyanide can be fatal.

Hospital Waste

Hospital waste is generated during the diagnosis, treatment, or immunization of human beings or animals or in research activities in these fields or in the production or testing of biologicals. It may include wastes like sharps, soiled waste, disposables, anatomical waste, cultures, discarded medicines, chemical wastes, etc. These are in the form of disposable syringes, swabs, bandages, body fluids, human excreta, etc. This waste is highly infectious and can be a serious threat to human health if not managed in a scientific and discriminate manner.

HIERARCHY OF SUSTAINABLE WASTE MANAGEMENT

The Hierarchy of Sustainable Waste Management **(Fig 6.1)** developed by the Earth Engineering Center at Columbia University is widely used as a reference to sustainable solid waste management and disposal. The hierarchy of waste management recognizes that reducing the use of materials and reusing them to be the most environmental friendly. Source reduction begins with reducing the amount of waste generated and reusing materials to prevent them from entering the waste stream. Thus, waste is not generated until the end of "reuse" phase. Once the waste is generated, it needs to be collected. Material recovery from waste in the form of recycling and composting is recognized to be the most effective way of handling wastes.

Source Reduction and Reuse

Source reduction, also known as waste prevention, means reducing waste at the source. It can take many different forms, including reusing or donating items, buying in bulk, reducing packaging, redesigning products, and reducing toxicity. Source reduction also is important in manufacturing. Light weighting of packaging, reuse, and remanufacturing are all becoming more popular business trends. Purchasing products that incorporate these features supports source reduction. Source reduction can:

(i) Save natural resources.

(ii) Conserve energy.

(iii) Reduce pollution.

(iv) Reduce the toxicity of our waste.

(v) Save money for consumers and businesses alike.

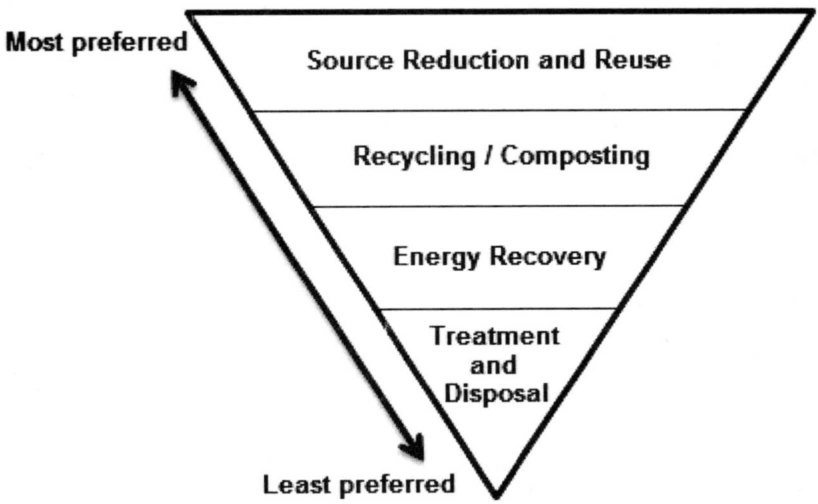

Fig. 6.1: Sustainable waste management hierarchy

Recycling/Composting

Recycling is a series of activities that includes the collection of used, reused, or unused items that would otherwise be considered waste; sorting and processing the recyclable products into raw materials; and remanufacturing the recycled raw materials into new products. Consumers provide the last link in recycling by purchasing products made from recycled content. Recycling also can include composting of food scraps, yard trimmings, and other organic materials. Recycling prevents the emission of many greenhouse gases and water pollutants, saves energy, supplies valuable raw materials to industry, creates jobs, stimulates the development of greener technologies, conserves resources for our children's future, and reduces the need for new landfills and combustors.

Energy Recovery

Energy recovery from waste is the conversion of non-recyclable waste materials into useable heat, electricity, or fuel through a variety of processes, including combustion, gasification, pyrolization, anaerobic digestion, and landfill gas (LFG) recovery. This process is often called waste-to-energy (WTE).

Treatment and Disposal

Landfills are the most common form of waste disposal and are an important component of an integrated waste management system. Landfills that accept municipal solid waste are primarily regulated by state, tribal, and local governments. Today's landfills must meet stringent design, operation, and closure requirements. Combustion of solid waste is done to reduce the amount of landfill space needed. Methane gas, a byproduct of decomposing waste, can be collected and used as fuel to generate electricity. After a landfill is capped, the land may be used for recreation sites such as parks, golf courses, and ski slopes.

SOLID WASTE MANAGEMENT

A solid waste management (SWM) system includes the generation of waste, storage, collection, transportation, processing and final disposal. Managing solid waste generally involves planning, financing, construction and operation of facilities for the collection, transportation, recycling and final disposition of the waste. Solid waste management (SWM) is a basic public necessity and this service is provided by respective urban local bodies. The waste management term usually relates to materials produced by human activity, and the process is generally undertaken to reduce their effect on health, the environment or aesthetics. Waste management is a distinct practice from resource recovery which focuses on delaying the rate of consumption of natural resources **(Fig 6.2)**. All wastes materials, whether they are solid, liquid, gaseous or radioactive fall within the remit of waste management.

Waste management practices can differ for developed and developing nations, for urban and rural areas, and for residential and industrial producers. Management of non-hazardous waste residential and institutional waste in metropolitan areas is usually the responsibility of local government authorities, while management for non-hazardous commercial and industrial waste is usually the responsibility of the generator subject to local, national or international controls.

Waste composition dictates the waste management strategy to be employed in a particular location. Organics are putrescible, and are food for pests and insects and hence need to be collected and disposed off on a daily basis. The amount of recyclables like paper and plastic dictates how often they need to be collected. Recyclables represent an immediate monetary value to the collectors. Organics need controlled biological treatment to be of any value, however due to the general absence of such facilities, organics do not represent any direct value to informal collectors.

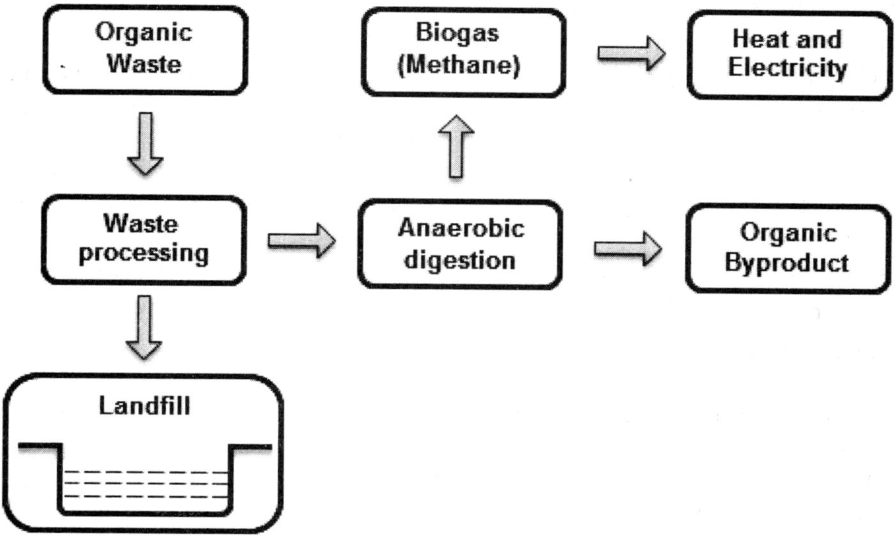

Fig. 6.2: Solid Waste Management

The municipal solid waste industry has four components: recycling, composting, landfilling, and waste-to-energy via incineration. The primary steps are generation, collection, sorting and separation, transfer, and disposal.

Collection

The functional element of collection includes not only the gathering of solid waste and recyclable materials, but also the transport of these materials, after collection, to the location where the collection vehicle is emptied. This location may be materials processing facility, a transfer station or a landfill disposal site. Waste transfer stations are facilities where municipal solid waste is unloaded from collection vehicles and briefly held while it is reloaded onto larger long-distance transport vehicles for shipment to landfills or other treatment or disposal facilities. By combining the loads of several individual waste collection trucks into a single shipment, communities can save money on the labor and operating costs of transporting the waste to a distant disposal site. They can also reduce the total number of vehicular trips traveling to and from the disposal site. Although waste transfer stations help reduce the impacts of trucks traveling to and from the disposal site, they can cause an increase in traffic in the immediate area where they are located. If not properly sited, designed and operated they can cause problems for residents living near them.

Disposal

Today, the disposal of wastes by land filling or land spreading is the ultimate fate of all solid wastes, whether they are residential wastes collected and transported directly to a landfill site, residual materials from materials recovery facilities (MRFs), residue from the combustion of solid waste, compost, or other substances from various solid waste processing facilities. As cities are growing in size with a rise in the population, the amount of waste generated is increasing becoming unmanageable. The local corporations have adapted different methods for the disposal of waste – open dumps, composting, landfills, sanitary landfills, and incineration plants.

Open Dumps

Open dumps refer to uncovered areas that are used to dump solid waste of all kinds. The waste is untreated, uncovered, and not segregated. It is the breeding ground for flies, rats, and other insects that spread disease. The rainwater run-off from these dumps contaminates nearby land and water thereby spreading disease. In some countries, open dumps are being phased out.

Landfills

Landfills are generally located in urban areas where a large amount of waste is generated and has to be dumped in a common place. Unlike an open dump, it is a pit that is dug in the ground. The garbage is dumped and the pit is covered thus preventing the breeding of flies and rats. At the end of each day, a layer of soil is scattered on top of it and some mechanism, usually earth-moving equipment is used to compress the garbage, which now forms a cell. Thus, every day, garbage is dumped and becomes a cell. After the landfill is full, the area is covered with a thick layer of mud and the site can thereafter be developed as a parking lot or a park. Landfills have many problems. All types of waste is dumped in landfills and when water seeps through them it gets contaminated and in turn pollutes the surrounding area. This contamination of groundwater and soil through landfills is known as leaching. Some landfills are also used for waste management purposes, such as the temporary storage, consolidate ion and transfer, or processing of waste material (sorting, treatment, or recycling).

Modern landfills **(Fig 6.3)** are well-engineered facilities that are located, designed, operated, and monitored to ensure compliance with environmental regulations. Solid waste landfills must be designed to protect the environment from contaminants which may be present in the solid waste stream. The landfill siting plan—which prevents the sitting of landfills in environmentally-sensitive areas—as well as on-site environmental monitoring systems—which monitor for any sign of groundwater contamination and for landfill gas—provide additional safeguards. In addition, many new landfills collect potentially harmful landfill gas emissions and convert the gas into energy.

Municipal solid waste landfills (MSWLFs) receive household waste. They can also receive non-hazardous sludge, industrial solid waste, and construction and demolition debris. All solid waste landfills must comply with the following regulations:

a) Location restrictions - ensure that landfills are built in suitable geological areas away from faults, wetlands, flood plains, or other restricted areas.

b) Composite liners requirements - include a flexible membrane (geomembrane) overlaying two feet of compacted clay soil lining the bottom and sides of the landfill, protect groundwater and the underlying soil from leachate releases.

c) Leachate collection and removal systems - sit on top of the composite liner and removes leachate from the landfill for treatment and disposal.

d) Operating practices - include compacting and covering waste frequently with several inches of soil help reduce odor; control litter, insects, and rodents; and protect public health.

e) Groundwater monitoring requirements - requires testing groundwater wells to determine whether waste materials have escaped from the landfill.

f) Closure and post closure care requirements - include covering landfills and providing long-term care of closed landfills.

g) Corrective action provisions - control and clean up landfill releases and achieve groundwater protection standards.

Some materials may be banned from disposal in municipal solid waste landfills including common household items such as paints, cleaners/chemicals, motor oil, batteries, and pesticides. Leftover portions of these products are called household hazardous waste. These products, if mishandled, can be dangerous to your health and the environment. Many municipal landfills have a household hazardous waste drop-off station for these materials.

Impacts of Landfills

Many adverse impacts may occur from landfill operations. Damage can include infrastructure disruption (e.g. damage to access roads by heavy vehicles); pollution of the local environment (such as contamination of groundwater and/or aquifers by leakage or sinkholes and residual soil contamination during landfill usage, as well as after landfill closure); off gassing of methane generated by decaying organic wastes (methane is a greenhouse gas many times more potent than carbon dioxide, and can itself be a danger to inhabitants of an area); harboring of disease vectors such as rats and flies, particularly from improperly operated landfills, which are common in developing countries; injuries to wildlife; and simple nuisance problems (e.g., dust, odor, vermin, or noise pollution).

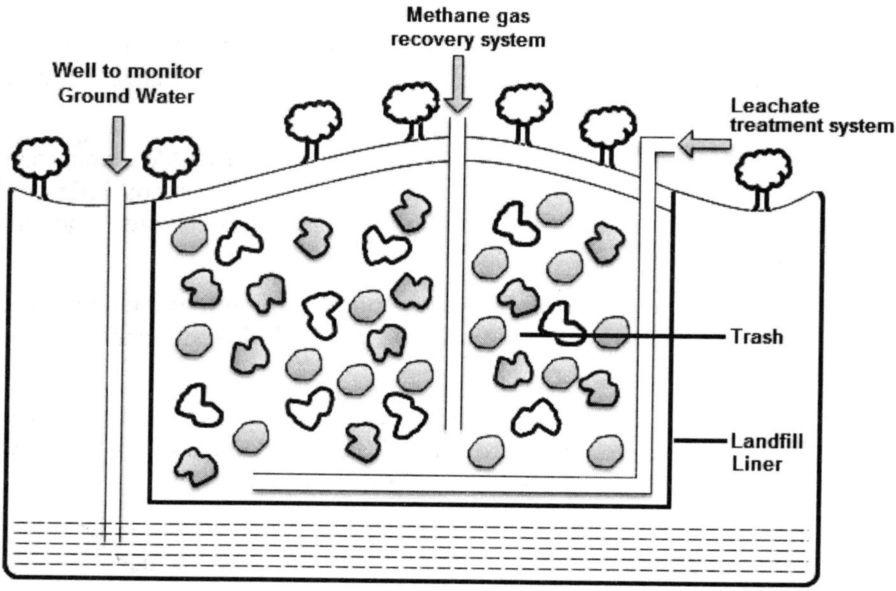

Fig. 6.3: Landfill

Sanitary Landfills

An alternative to landfills which will solve the problem of leaching to some extent is a sanitary landfill which is more hygienic and built in a methodical manner. These are lined with materials that are impermeable such as plastics and clay, and are also built over impermeable soil. Constructing sanitary landfills is very costly and they are having their own problems. Some authorities claim that often the plastic liner develops cracks as it reacts with various chemical solvents present in the waste.

The rate of decomposition in sanitary landfills is also extremely variable. This can be due to the fact that less oxygen is available as the garbage is compressed very tightly. It has also been observed that some biodegradable materials do not decompose in a landfill. Another major problem is the development of methane gas, which occurs when little oxygen is present, i.e. during anaerobic decomposition. In some countries, the methane being produced from sanitary landfills is tapped and sold as fuel.

Landfill Mining

Landfill mining and reclamation (LFMR) is a process whereby solid wastes which have previously been land filled are excavated and processed. The function of landfill mining is to reduce the amount of landfill mass encapsulated within the closed landfill and/or temporarily remove hazardous material to allow protective measures to be taken before the landfill mass is replaced. In the process, mining recovers valuable recyclable materials, a combustible fraction, soil, and landfill space. The aeration of the landfill soil is a secondary benefit regarding the landfill's future use. The combustible fraction is useful for the generation of power. The overall appearance of the landfill mining procedure is a sequence of processing machines laid out in a functional conveyor system. The operating principle is to excavate, sieve and sort the landfill material. Landfill mining is also possible in countries where land is not available for new landfill sites. In this instance landfill space can be reclaimed by the extraction of biodegradable waste and other substances then refilled with wastes requiring disposal.

Bioreactor Landfill

A sanitary landfill operated for the purpose of rapid stabilization of the decomposable organic waste constituents by purposeful control of biological processes is termed as bioreactor landfill. It has many benefits like:

1) Increase disposal capacity.
2) Provides flexibility in leachate.
3) Enhances feasibility of landfill gas to energy projects etc.

There are three different general types of bioreactor landfill configurations:

1) **Aerobic**: In an aerobic bioreactor landfill **(Fig 6.4)**, leachate is removed from the bottom layer, piped to liquids storage tanks, and re-circulated into the landfill in a controlled manner. Air is injected into the waste mass, using vertical or horizontal wells, to promote aerobic activity and accelerate waste stabilization.
2) **Anaerobic**: In an anaerobic bioreactor landfill, moisture is added to the waste mass in the form of re-circulated leachate and other sources to obtain optimal moisture levels. Biodegradation occurs in the absence of oxygen (anaerobically) and produces landfill gas. Landfill gas, primarily methane, can be captured to minimize greenhouse gas emissions and for energy projects.
3) **Hybrid (Aerobic-Anaerobic):** The hybrid bioreactor landfill accelerates waste degradation by employing a sequential aerobic-anaerobic treatment to rapidly degrade organics in the upper sections of the landfill and collect gas from lower sections. Operation as a hybrid results in the earlier onset of methanogenesis compared to aerobic landfills.

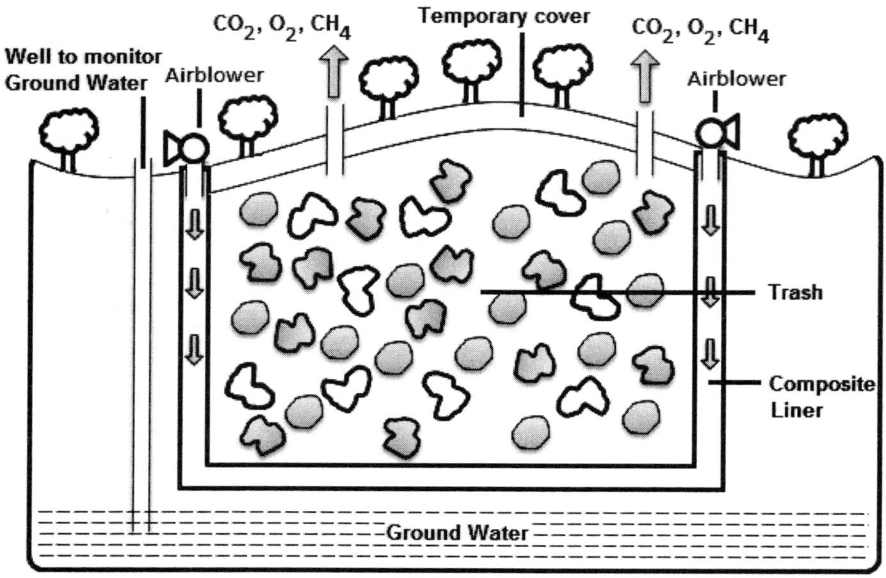

Fig. 6.4: Aerobic Landfill Bioreactor

Potential advantages of bioreactors include:

(i) Decomposition and biological stabilization in years vs. decades in "dry tombs".

(ii) Lower waste toxicity and mobility due to both aerobic and anaerobic conditions.

(iii) Reduced leachate disposal costs.

(iv) A 15 to 30 percent gain in landfill space due to an increase in density of waste mass.

(v) Significant increased LFG generation that, when captured, can be used for energy use onsite or sold.

(vi) Reduced post-closure care.

Research has shown that municipal solid waste can be rapidly degraded and made less hazardous (due to degradation of organics and the sequestration of inorganics) by enhancing and controlling the moisture within the landfill under aerobic and/or anaerobic conditions. Leachate quality in a bioreactor rapidly improves which leads to reduced leachate disposal costs.

Incineration

The process of burning waste in large furnaces is known as incineration. In these plants the recyclable material is segregated and the rest of the material is burnt. At the end of the process all that is left behind is ash. During the process some of the ash floats out with the hot air. This is called fly ash. Both the fly ash and the ash that is left in the furnace after burning have high concentrations of dangerous toxins such as dioxins and heavy metals. Disposing of this ash is a problem. The ash that is buried at the landfills leaches the area and cause severe contamination.

Burning garbage is not a clean process as it produces tonnes of toxic ash and pollutes the air and water. A large amount of the waste that is burnt here can be recovered and recycled. In fact, at present, incineration is kept as the last resort and is used mainly for treating the infectious waste.

Composting

Composting is a biological process in which micro-organisms, mainly fungi and bacteria, convert degradable organic waste into humus like substance. This finished product, which looks like soil, is high in carbon and nitrogen and is an excellent medium for growing plants. The process of composting ensures the waste that is produced in the kitchens is not carelessly thrown and left to rot. It recycles the nutrients and returns them to the soil as nutrients. Apart from being clean, cheap, and safe, composting can significantly reduce the amount of disposable garbage. The organic fertilizer can be used instead of chemical fertilizers and is better specially when used for vegetables. It increases the soil's ability to hold water and makes the soil easier to cultivate. It helped the soil retain more of the plant nutrients.

Compost is organic material that can be used as a soil amendment or as a medium to grow plants. Mature compost is a stable material with content called humus that is dark brown or black and has a soil-like, earthy smell. It is created by: combining organic wastes (e.g., yard trimmings, food wastes, manures) in proper ratios into piles, rows, or vessels; adding bulking agents (e.g., wood chips) as necessary to accelerate the breakdown of organic materials; and allowing the finished material to fully stabilize and mature through a curing process. Some important benefits of composting are as follows:

a) Compost allows the soil to retain more plant nutrients over a longer period.

b) It supplies part of the 16 essential elements needed by the plants.

c) It helps reduce the adverse effects of excessive alkalinity, acidity, or the excessive use of chemical fertilizer.

d) It makes soil easier to cultivate.

e) It helps keep the soil cool in summer and warm in winter.

f) It aids in preventing soil erosion by keeping the soil covered.

g) It helps in controlling the growth of weeds in the garden.

h) Facilitate reforestation, wetlands restoration, and habitat revitalization efforts by amending contaminated, compacted, and marginal soils.

i) Cost-effectively remediate soils contaminated by hazardous waste.

j) Remove solids, oil, grease, and heavy metals from storm water runoff.

k) Avoids Methane and leachate formulation in landfills.

Factors Affecting Composting Process

Feedstock and nutrient balance
Controlled decomposition requires a proper balance of "green" organic materials (e.g., grass clippings, food scraps, manure), which contain large amounts of nitrogen, and "brown" organic materials (e.g., dry leaves, wood chips, branches), which contain large amounts of carbon but little nitrogen. Obtaining the right nutrient mix requires experimentation and patience and is part of the art and science of composting.

Particle size
Grinding, chipping, and shredding materials increases the surface area on which the microorganism can feed. Smaller particles also produce a more homogeneous compost mixture and improve pile insulation to help maintain optimum temperatures. If the particles are too small, however, they might prevent air from flowing freely through the pile.

Moisture content
Microorganisms living in a compost pile need an adequate amount of moisture to survive. Water is the key element that helps transports substances within the compost pile and makes the nutrients in organic material accessible to the microbes. Organic material contains some moisture in varying amounts, but moisture also might come in the form of rainfall or intentional watering.

Oxygen flow
Turning the pile, placing the pile on a series of pipes, or including bulking agents such as wood chips and shredded newspaper all help aerate the pile. Aerating the pile allows decomposition to occur at a faster rate than anaerobic conditions. Care must be taken, however, not to provide too much oxygen, which can dry out the pile and impede the composting process.

Temperature
Microorganisms require a certain temperature range for optimal activity. Certain temperatures promote rapid composting and destroy pathogens and weed seeds. Microbial activity can raise the temperature of the pile's core to at least 140° F. If the temperature does not increase, anaerobic conditions

(i.e., rotting) occur. Controlling the previous four factors can bring about the proper temperature.

Types of Composting

Backyard or Onsite Composting

Backyard or onsite composting can be conducted by residents and other small-quantity generators of organic waste on their own property. By composting these materials onsite, homeowners and select businesses can significantly reduce the amount of waste that needs to be disposed of and thereby save money from avoided disposal costs.

Backyard or onsite composting is suitable for converting yard trimmings and food scraps into compost that can be applied on site. This method should not be used to compost animal products or large quantities of food scraps. Households, commercial establishments, and institutions (e.g., universities, schools, and hospitals) can leave grass clippings on the lawn-known as "grass cycling"-where the cuttings will decompose naturally and return some nutrients back to the soil. Backyard or onsite composters also might keep leaves in piles for eventual use as mulch around trees and scrubs to retain moisture.

Climate and seasonal variations do not present major challenges to backyard or onsite composting because this method typically involves small quantities of organic waste. When conditions change-for example, if a rainy season approaches-the process can be adjusted accordingly without many complications.

Aerated (Turned) Windrow Composting

Organic waste is formed into rows of long piles called "windrows" and aerated by turning the pile periodically by either manual or mechanical means. The ideal pile height, which is between 4 and 8 feet, allows for a pile large enough to generate sufficient heat and maintain temperatures, yet small enough to allow oxygen to flow to the windrow's core **(Fig 6.5)**. The ideal pile width is between 14 and 16 feet. This method can accommodate large volumes of diverse wastes, including yard trimmings, grease, liquids, and animal byproducts (such as fish and poultry wastes), but only with frequent turning and careful monitoring. This method is suited for large quantities, such as that generated by entire communities and collected by local governments, and high volume food-processing businesses (e.g., restaurants, cafeterias, packing plants).

In a warm, arid climate, windrows are sometimes covered or placed under a shelter to prevent water from evaporating. In rainy seasons, the shapes of the pile can be adjusted so that water runs off the top of the pile rather than being absorbed into the pile. Also, windrow composting can work in cold climates. Often the outside of the pile might freeze, but in its core, a windrow can reach 140° F. Leachate is liquid released during the composting process. This can contaminate local ground-water and surface-water supplies and should be collected and treated.

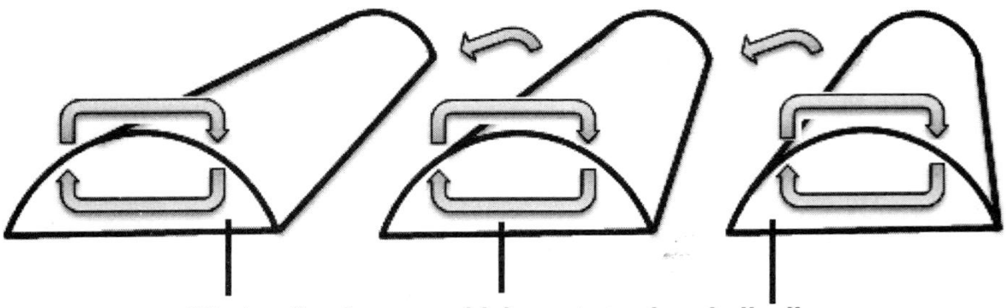

Fig. 6.5: Aerated windrow composting

In addition, windrow composting is a large scale operation and might be subject to regulatory enforcement. Samples of the compost should be tested in a laboratory for bacterial and heavy metal content. Odors also need to be controlled. The public should be informed of the operation and have a method to address any complaints about animals or bad odors. Other concerns might include zoning and sitting requirements. Windrow composting often requires large tracts of land, sturdy equipment, a continual supply of labor to maintain and operate the facility, and patience to experiment with various materials mixtures and turning frequencies. This method will yield significant amounts of compost, which might require assistance to market the end-product. Alternatively, local governments can make the compost available to residents for a low or no cost.

Aerated Static Pile Composting

In aerated static pile composting **(Fig 6.6)**, organic waste is mixed together in one large pile instead of rows. To aerate the pile, layers of loosely piled bulking agents (e.g., wood chips, shredded newspaper) are added so that air can pass from the bottom to the top of the pile. The piles also can be placed over a network of pipes that deliver air into or draw air out of the pile. Air blowers might be activated by a timer or temperature sensors. Aerated static piles are suitable for a relatively homogenous mix of organic waste and work well for larger quantity generators of yard trimmings and compostable municipal solid waste (e.g., food scraps, paper products), which might include local governments, landscapers, or farms. This method, however, does not work well for composting animal byproducts or grease from food processing industries.

Like windrow composting, in a warm, arid climate, aerated static piles are sometimes covered or placed under a shelter to prevent water from evaporating. In the cold, the core of the pile will retain its warm temperature, but aeration might be more difficult in the cold because this method involves passive air flowing rather than active turning. Some aerated static piles are placed indoors with proper ventilation.

Since there is no physical turning, this method requires careful monitoring to ensure that the outside of the pile heats up as much as the core. One way to alleviate bad odors is to apply a thick layer of finished compost over the pile, which can help maintain high temperatures throughout the pile. Another way to deal with odor, provided that the air blower draws air out of the pile, is to filter this air through a biofilter made from finished compost. This method typically requires equipment such as blowers, pipes, sensors, and fans, which might involve significant costs and technical assistance. Having a controlled supply of air enables construction of large piles, which require less land than the windrow method.

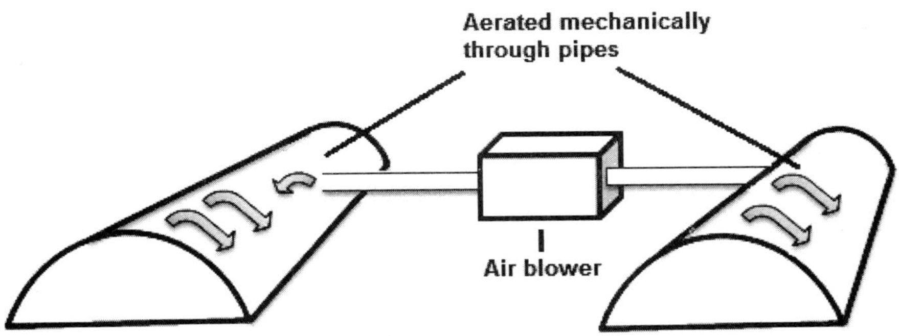

Fig. 6.6: Aerated static pile composting

In-Vessel Composting

Organic materials are fed into a drum, silo, concrete-lined trench, or similar equipment where the environmental conditions-including temperature, moisture, and aeration-are closely controlled. The apparatus usually has a mechanism to turn or agitate the material for proper aeration. In-vessel composters vary in size and capacity. In-vessel composting can process large amounts of waste without taking up as much space as the windrow method. In addition, it can accommodate virtually any type of organic waste (e.g., meat, animal manure, biosolids, food scraps). Some in-vessel composters can fit into a school or restaurant kitchen while others can be as large as a school bus to accommodate large food processing plants.

In-vessel composting **(Fig 6.7)** can be used year-round in virtually any climate because the environment is carefully controlled, often by electronic means. This method can even be used in extremely cold weather if the equipment is insulated or the processing takes place indoors. In-vessel composting produces very little odor and minimal leachate. In-vessel composters are expensive and might require technical assistance to operate properly, but this method uses much less land and manual labor than windrow composting. Conversion of organic material to compost can take as little as a few weeks. Once the compost comes out of the vessel, however, it still requires a few more weeks or months for the microbial activity to stabilize and the pile to cool.

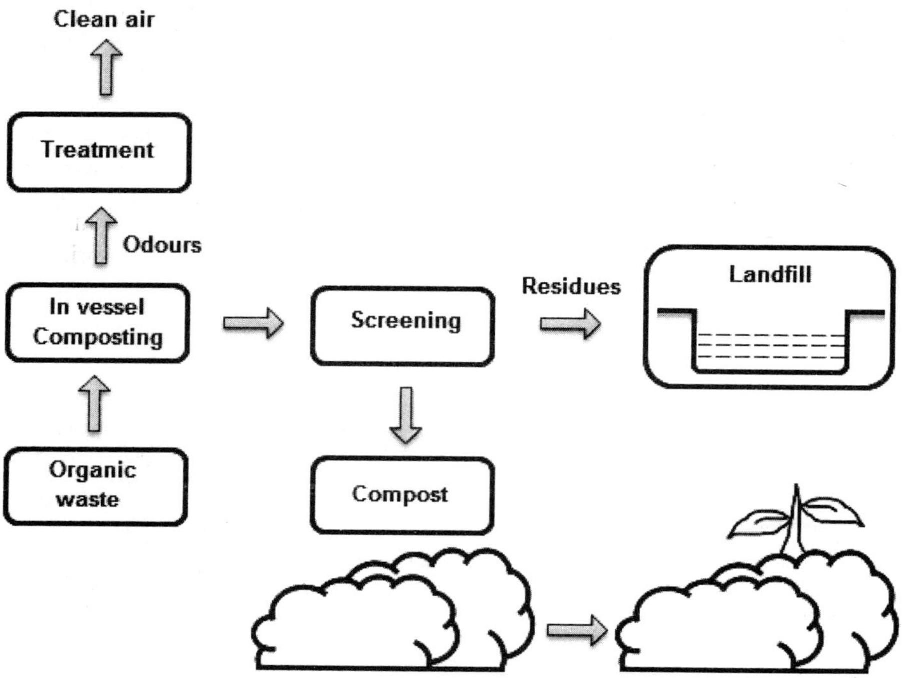

Fig. 6.7: In-vessel composting

Vermicomposting

Vermicompost is the product or process of composting using various worms, usually red wigglers, white worms, and other earthworms to create a heterogeneous mixture of decomposing vegetable or food waste, bedding materials, and vermicast. Vermicast, also called worm castings, worm humus or worm manure, is the end-product of the breakdown of organic matter by an earthworm. These castings have been shown to contain reduced levels of contaminants and a higher saturation of nutrients than do organic materials before vermicomposting. Containing water-soluble nutrients, vermicompost is an excellent, nutrient-rich organic fertilizer and soil conditioner. This process of producing vermicompost is called vermicomposting.

Worm Species used in vermicomposting

One of the earthworm species most often used for composting is the Red Wiggler (*Eisenia fetida* or *Eisenia andrei*); *Lumbricus rubellus* is another breed of worm that can be used, but it does not adapt as well to the shallow compost bin as does *Eisenia fetida*. European night crawlers (*Eisenia hortensis*) may also be used. African Night crawlers (*Eudrilus eugeniae*) are another set of popular composters. Blueworms (*Perionyx excavatus*) may be used in the tropics.

Worms will eat almost anything you would put in a typical compost pile (e.g., food scraps, paper, and plants). Vermicomposting can be ideal for apartment dwellers or small offices that want to derive some of the benefits of composting and reduce solid waste. It is frequently used in schools to teach children conservation and recycling.

Worms are sensitive to variations in climate. Extreme temperatures and direct sunlight are not healthy for the worms. The optimal temperatures for vermicomposting range from 55° F to 77° F. In hot, arid areas, the bin should be placed under the shade. By vermicomposting indoors, however, one can avoid many of the problems posed by hot or cold climates. The primary responsibility is to keep the worms alive and healthy by providing the proper conditions and sufficient food.

Benefits of vermicomposting

a) It improves Soil aeration and enriches soil with micro-organisms (adding enzymes such as phosphatase and cellulase).

b) Microbial activity in worm castings is 10 to 20 times higher than in the soil and organic matter that the worm ingests.

c) Attracts deep-burrowing earthworms already present in the soil.

d) Enhances germination, plant growth, and crop yield.

e) Biowastes conversion reduces waste flow to landfills.

f) Production reduces greenhouse gas emissions such as methane and nitric oxide produced in landfills or incinerators when not composted or through methane harvest.

g) Vermicompost can be mixed directly into the soil, or steeped in water and made into a worm tea by mixing some vermicompost in water, bubbling in oxygen with a small air pump, and steeping for a number of hours or days, which can be used as fertilizer.

REFERENCES

Ashare, E. D. Wise and R. Wentworth, 1977 Fuel Gas Recovery from Animal Residue, Dynatech R/D, Available from the US National Technical Information Service (NTIS).

Asnani, P.U. (2004). United States Asia Environmental Partnership Report, United States Agency for International Development, Centre for Environmental Planning and Technology, Ahmedabad.

Augenstein, D. 1992 The Greenhouse Effect and US Landfill Methane. Global Environmental Change, p. 311-328 December.

Augenstein, D., R. Yazdani, K. Dahl and R. Moore, 1998, Yolo County Controlled Landfill Project. Proceedings, California Integrated Waste Management Board (CIWMB) Landfill Gas Assessment and Management Symposium, Ontario, CA April. CIWMB, Sacramento, CA.

Augenstein, D. R. Yazdani, L. Sinderson, J. Kieffer M. Byars, 2000 Proceedings, Second International Methane Mitigation Symposium, US EPA and Russian Academy of Sciences (Siberian Branch) Akademgorodok, Novosibirsk, Siberia, Russia. Available from EPA, Washington DC).

Jain, A.P. and G.B. Pant, 1994. Solid Waste Management in India. Conference paper presented at the 20th WEDC Conference, Colombo, Sri Lanka.

Japan Waste Management Association, 1996. Waste Management in Japan 1996. Tokyo, Japan.

Johannessen, L.M., 1998. Technical Report on Asia: The Emerging Approach to Landfilling of Municipal Solid Waste. Transportation, Water and Urban Development Department, World Bank, Washington, D.C., USA.

Listyawan, B., 1996. Prospects of Recycling Systems in Indonesia. Recycling in Asia: Partnerships for Responsive Solid Waste Management. United Nations Centre for Regional Development (UNCRD), Nagoya, Japan.

MacFarlane, C., 1998. Solid Waste Management Consultant, Markham, Canada (personal communication).

McDonald's Corporation, 1997. The Annual: McDonald's Corporation 1996 Annual Report. Oak Brook, Illinois, USA.

McGee, T.G. and C.J. Griffiths, 1993. Global Urbanization: Towards the Twenty-First Century. Population Distribution and Migration. Draft proceedings of the United Nations Expert Meeting on Population Distribution and Migration, Santa Cruz, Bolivia, January 18—22 (United Nations, New York, August, 1994).

National Packaging Monitoring System, 1993. Database accessible through the Solid Waste Management Division, Environment Canada, Ottawa. Cited in Environment Canada, 1998. State of the Environment Fact Sheet No. 95—1.

Organisation for Economic Co-operation and Development (OECD), 1991. Environmental Labeling in OECD Countries. Publications Service, Paris, France.

Organisation for Economic and Co-operation and Development (OECD), 1995. OECD Environmental Data: Compendium 1995. Publications Service, Paris, France.

OECD (1995).OECD Environmental Data Compendium 1995, OECD, Paris.

Parikh, J. et al., 1991. Consumption Patterns: The Driving Force of Environmental Stress.

Perla, M., 1997. Community Composting in Developing Countries. Biocycle, June, pp. 48–51.

Planning, Environment and Lands Bureau, 1998. Hong Kong Special Administration of the People's Republic of China. Website. http://www.pelb.wpelb.gov.hk/waste/current.htm

Pollution Control Department, 1998 (personal communication with staff, Bangkok, Thailand).

Porter, R., 1996. The Economics of Water and Waste: A Case Study of Jakarta, Indonesia. Avebury Ashgate Publishing Ltd., England.

Powell, J., 1983. A Comparison of the Energy Savings from the Use of Secondary Materials. Conservation & Recycling 6 (1/2), pp. 27—32. Cited in van Beukering, P., 1994. An Economic Analysis of Different Types of Formal and Informal Entrepreneurs, Recovering Urban Solid Waste in Bangalore (India). Resources, Conservation and Recycling (12), pp. 229—252.

NEERI (1995). 'Strategy Paper on SWM in India', National Environmental Engineering Research Institute, Nagpur.

SC (1999). Report of the Supreme Court Appointed Committee on Solid Waste Management in Class I Cities in India, Supreme Court of India, New Delhi.

7

Global Environmental Problems

A variety of environmental problems now affect our entire world. Every environmental problem has causes, numerous effects, and most importantly, a solution. Some of the largest problems now affecting the world are Acid Rain, Air Pollution, Global Warming, Hazardous Waste, Ozone Depletion, Smog, Water Pollution, Overpopulation, and Rain Forest Destruction. The climate is changing. The earth is warming up, and there is now overwhelming scientific consensus that it is happening, and human-induced. With global warming on the increase and species and their habitats on the decrease, chances for ecosystems to adapt naturally are diminishing. Many are agreed that climate change may be one of the greatest threats facing the planet. Recent years show increasing temperatures in various regions, and/or increasing extremities in weather patterns.

Global warming and climate change refer to an increase in average global temperatures. Natural events and human activities are believed to be contributing to an increase in average global temperatures. This is caused primarily by increases in "greenhouse" gases such as Carbon Dioxide (CO2). A warming planet thus leads to a change in climate which can affect weather in various ways. Pollution is presence of matter (gas, liquid, solid) or energy (heat, noise, radiation) whose nature, location, or quantity directly or indirectly alters characteristics or processes of any part of the environment, and causes (or has the potential to cause) damage to the condition, health, safety, or welfare of animals, humans, plants, or property. Pollution is a growing pain. Pollution is not a problem that came suddenly from the sky; it's our fault and has been a part of our life through many years. We must be wise in managing our resources, and take positive action towards preventing any forms of pollution to the environment. Make the world a better place to live.

Measurements of temperature taken by instruments all over the world, on land and at sea have revealed that during the 20th century the Earth's surface and lowest part of the atmosphere warmed up on average by about 0.6°C. During this period, man-made emissions of greenhouse gases, including carbon dioxide, methane and nitrous oxide have increased, largely as a result of the burning of fossil fuels for energy and transportation, and land use changes including deforestation for agriculture. In the last 20 years, concern has grown that these two phenomena are, at least in part, associated with each other. That is to say, global warming is now considered most probably to be due to the increases in greenhouse gas emissions and concurrent increases in atmospheric greenhouse gas concentrations, which have enhanced the Earth's natural greenhouse effect.

In addition to the natural fluxes of carbon through the Earth system, anthropogenic (human) activities, particularly fossil fuel burning and deforestation, are also releasing carbon dioxide into the atmosphere. When we mine coal and extract oil from the Earth's crust, and then burn these fossil fuels for transportation, heating, cooking, electricity, and manufacturing, we are effectively moving carbon more rapidly into the atmosphere than is being removed naturally through the sedimentation of carbon, ultimately causing atmospheric carbon dioxide concentrations to increase. Also, by clearing forests to support agriculture, we are transferring carbon from living biomass into the atmosphere (dry wood is about 50 percent carbon). The result is that humans are adding ever-increasing amounts of extra carbon dioxide into the atmosphere. Because of this, atmospheric carbon dioxide concentrations are higher today than they have been over the last half-million years or longer. Today, global warming is the most serious problem among the problems which caused the increase in environmental temperature, climate change, rising sea surface water, ecological changes that give greater influence to the basic human existence. In addition, the problem of damage to the ozone layer, acid rain, green house gases and green house effect, and others to give effect to the health and the environment, not only environmental problems air, but also environmental problems of water and land which is in the condition which cannot be ignored. The major environmental issues world is facing today are:

1) Green house effect and Global warming
2) Acid rains
3) Ozone depletion
4) Radioactive waste

The climate changes that will result from global warming are extremely difficult to predict. The weather is determined by so many factors that it is often compared to chaos by scientists. Changing the temperature will likely have some effect on the planet's weather, but just what that effect will be is nearly impossible to predict. If temperatures do indeed rise significantly, the most important result would be that some portion of the polar icecaps would melt, raising global sea levels. The rise in sea levels would be disastrous for some places. Islands would disappear; meaning their millions of inhabitants would have to relocate. Flooding would occur along coastlines all over the world, displacing more people and ruining cropland. In the case of major global warming and melted ice caps, some countries might simply cease to exist. Global warming, if uncontrolled, could cause a major catastrophe.

Environmental problems can result in destruction of valuable natural environments such as forests, rivers, beaches and others, in addition to damage to the biological diversity that is essential for humans. Therefore there is need for international efforts to tackle this.

GREEN HOUSE EFFECT

Climate is the average or overall weather pattern of an area over an extended period of time. Climate is impacted by latitude (which determines the amount and directness of solar radiation that hits a particular area), wind currents, ocean currents, and topography. The greenhouse effect is the means by which the Earth retains heat and stays warm. The Earth's troposphere contains some small amounts of carbon dioxide, water, methane, ozone, chlorofluorocarbons (CFCs), and other greenhouse gases. When solar radiation passes through the atmosphere, these gases trap some of this radiation near the Earth, keeping the Earth warm. However, as humans produce increasing amounts of greenhouse gases and simultaneously decrease forest cover (which stores carbon dioxide and prevents it from reaching the atmosphere), there is increase in amount of heat trapped on the Earth's surface and thus raise the average temperature **(Fig 7.1)**. Researchers have predicted that the Earth's average temperature will rise to 3–5° C within the next 50 years. By comparison, the last Ice Age was only about 5° C cooler on average than today's climate.

Many of these greenhouse gases are actually life-enabling, for without them, heat would escape back into space and the Earth's average temperature would be a lot colder. However, if the greenhouse effect becomes stronger, then more heat gets trapped than needed, and the Earth might become less habitable for humans, plants and animals. Carbon dioxide, though not the most potent of greenhouse gases, is the most significant one. Human activity has caused an imbalance in the natural cycle of the greenhouse effect and related processes.

Causes of Green House Effect

Deforestation

One of the man-made causes of the greenhouse effect is deforestation. Deforestation increases the amount of carbon dioxide in the atmosphere. Also, due to the disappearance of trees, photosynthesis cannot take place. Deforestation is rampant today due to the burden of our needs on land. The levels of deforestation have increased by about 9% in recent times. Also, the burning of wood causes it to decay, therefore releasing more carbon-dioxide into the atmosphere.

Burning of Fossils

Greenhouse gases can also be released into the atmosphere due to the burning of fossil fuels, oil, coal and gas. These materials are used increasingly and rampantly in industries. Most factories also produce many gases which last for a longer time in the atmosphere. These gases contribute to the greenhouse effect and also increase the global warming on the planet. These gases are not naturally available in the atmosphere. Therefore industries are also a major cause of the greenhouse effect.

Electrical Appliances

Other man-made causes of the increase in the greenhouse effect are the emission of greenhouse gases by electrical appliances. Even the humble refrigerator in the house emits gases which contribute to the greenhouse effect. These gases are known as Chlorofluorocarbons (CFCs) and are used in refrigerators, aerosol cans, some foaming agents in the packaging industry, fire extinguisher chemicals, and cleaners used in the electronic industry. Some processes of the cement manufacturing industries also act as a cause towards the greenhouse effect.

Population Growth

Population growth is an indirect contributor and one of the major causes of the greenhouse effect. With the increase in population, the needs and wants of people increase. This increases the manufacturing and the industry process. This results in the increase of the release of industrial gases which catalyze the greenhouse effect.

Green House Gases

A greenhouse gas (sometimes abbreviated GHG) is a gas in an atmosphere that absorbs and emits radiation within the thermal infrared range. This process is the fundamental cause of the greenhouse effect. The primary greenhouse gases in the Earth's atmosphere are water vapor, carbon dioxide, methane, nitrous oxide, and ozone. Other than these, the list also includes chloro fluorocarbons (CFCs), hydro fluorocarbons (HFCs), perfluorocarbons (PFCs), sulfur hexafluoride, nitrogen trifluoride etc. As these gases are present in the atmosphere in small quantities, they are not as popular as carbon dioxide or methane. The major non-gas contributors to the Earth's greenhouse effect clouds, also absorbs and emits infrared radiation and thus have an effect on radiative properties of the greenhouse gases. Clouds are water droplets or ice crystals suspended in the atmosphere.

Atmospheric concentrations of greenhouse gases are determined by the balance between sources (emissions of the gas from human activities and natural systems) and sinks (the removal of the gas from the atmosphere by conversion to a different chemical compound). The proportion of an emission remaining in the atmosphere after a specified time is the "Airborne fraction" (AF). More precisely, the annual AF is the ratio of the atmospheric increase in a given year to that year's total emissions. For CO_2 the AF over the last 50 years (1956–2006) has been increasing at $0.25 \pm 0.21\%$/year.

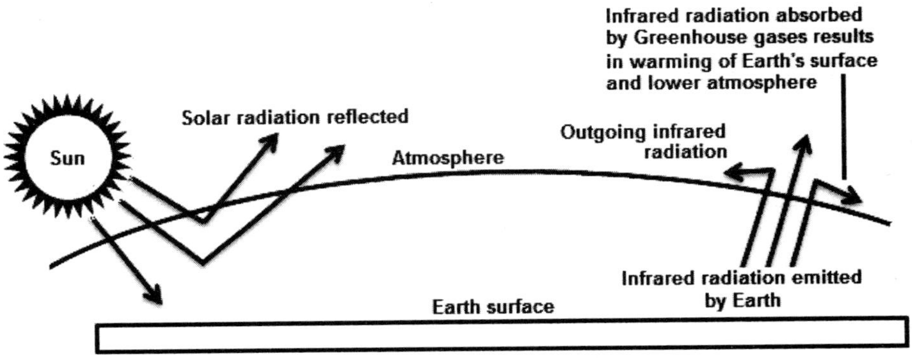

Fig. 7.1: Green House Effect

As for the greenhouse gases being portrayed in bad light today, it is primarily because of their contribution to the global warming phenomenon. Most of these gases are notorious for their global warming potential -- an estimate of how much a given mass of any greenhouse gas will contribute to the hazards of global warming. Abbreviated as GWP, it is determined by the efficiency of the molecule and the atmospheric lifetime of the said greenhouse gas. The atmospheric lifetime of a greenhouse gas is the estimated average time it will remain in the atmosphere after it has been emitted. The majorities of greenhouse gases arises from natural sources but are also contributed to by human activity. The major greenhouse gases with anthropogenic origins are described below:

Water Vapor

Water vapor, i.e., the gaseous form of water, which is produced as a result of evaporation of water and/or sublimation of ice, accounts for approximately 33 to 66 percent of the greenhouse gases in the atmosphere. The concentration of atmospheric water vapor across the globe is uneven, and therefore, it is difficult to determine its global warming potential. Anthropogenic factors, i.e., human activities, do contribute in the formation of water vapor, but the amount of vapor produced in this case is as good as none.

Carbon Dioxide

Carbon dioxide features second in the greenhouse gases list, accounting for 9 to 26 percent of total composition. While water vapor is primarily traced to the natural process of evaporation, carbon dioxide is released in air as a result of numerous anthropogenic activities; combustion of fossil fuels being the most important of the lot. As of May 2013, the concentration of carbon dioxide in the Earth's atmosphere has reached 400 ppm (parts per million) by volume, which is a significant rise from 280 ppm in pre-industrial times.

Methane

Methane accounts for anywhere between 4 to 9 percent of the greenhouse gases, but being highly potent, it is a bigger threat to the planet than carbon dioxide. In fact, its capacity of trapping heat is 25 times that of carbon dioxide. Methane is found in abundance beneath the Earth's crust from where it is released during the process of mining. Studies reveal that the amount of methane in the atmosphere has gone up from 700 parts per billion (ppb) in 1750 to 1,818 ppb in 2011.

Nitrous Oxide

Nitrous oxide is a naturally occurring regulator of the stratospheric ozone, which reacts with the ozone and contributes to the greenhouse effect. Averaged over a century, the impact of nitrous oxide per unit weight is 298 times that of carbon dioxide, which makes it a threat for the environment despite its low concentration. Agriculture is the main source of human-produced N_2O: cultivating soil, the use of nitrogen fertilizers, and animal waste handling can each stimulate naturally occurring bacteria to produce more N_2O. The livestock sector (primarily cows, chickens, and pigs) produces the majority of human-related N_2O. Industrial sources make up only about 20% of all anthropogenic sources, and include the production of nylon and nitric acid, and the burning of fossil fuel in internal combustion engines. Despite its relatively small concentration in the atmosphere, N_2O is the third largest greenhouse gas contributor to overall global warming, behind CO_2 and CH_4. (The other NO_x gases contribute to global warming indirectly, by contributing to troposphere ozone production during smog formation).

Ozone

Constituting approximately 3 to 7 percent of the total greenhouse gases, ozone acts as a greenhouse gas in the upper troposphere, where it absorbs the infrared energy that is emitted by the Earth. As with water vapor, even the concentration of ozone is uneven, which, in turn, makes it difficult to determine its global warming potential? However, it's believed that the radiative forcing of ozone present in troposphere is approximately 25 percent more than that of carbon dioxide.

Hydrofluorocarbons (HFC)

HFCs are man-made chemicals, many of which have been developed as alternatives to ozone-depleting substances for industrial, commercial, and consumer products. GWPs of HFCs range from 140 (HFC-152a) to 11,700 (HFC-23). The atmospheric lifetimes for HFCs vary from just over a year for HFC-152a to 260 years for HFC-23. Most commercial HFCs have atmospheric lifetimes under 15 years; eg. HFC-134a (used in vehicle air conditioning and refrigeration), has an atmospheric life of 14 years.

Perfluorocarbons (PFC)

PFCs are commonly used in refrigerating units and "clean" fire extinguishers. However, PFCs are extremely potent greenhouse gases, and can persist for up to 50,000 years. PFCs have extremely stable molecular structures and are largely immune to chemical processes in the lower atmosphere that break down most atmospheric pollutants. Not until the PFCs reach the mesosphere, about 60 kilometers above Earth, do very high-energy ultraviolet rays from the sun destroy them. This removal mechanism is extremely slow and as a result PFCs accumulate in the atmosphere and persist for several thousand years. The estimated atmospheric lifetimes for CF 4 and C 2F 6 are 50,000 and 10,000 years respectively.

Effects of Global Warming

Global warming is the long-term, cumulative effect that greenhouse gases, primarily carbon dioxide and methane, have on Earth's temperature when they build up in the atmosphere and trap the sun's heat. The planet is warming, from North Pole to South Pole, and everywhere in between. Globally, the mercury is already up more than 1 degree Fahrenheit (0.8 degree Celsius), and even more in sensitive Polar Regions. And the effects of rising temperatures aren't waiting for some far-flung future. They're happening right now. Signs are appearing all over, and some of them are surprising. The heat is not only melting glaciers and sea ice; it's also shifting precipitation patterns and setting animals on the move.

Some impacts from increasing temperatures are already happening:

1) Ice is melting worldwide, especially at the Earth's poles. This includes mountain glaciers, ice sheets covering West Antarctica and Greenland, and Arctic sea ice.

2) Researchers have tracked the decline of the penguins on Antarctica, where their numbers have fallen from 32,000 breeding pairs to 11,000 in 30 years.

3) Sea level rise became faster over the last century.

4) Some butterflies, foxes, and alpine plants have moved farther north or to higher, cooler areas.

5) Precipitation (rain and snowfall) has increased across the globe, on average.

6) Spruce bark beetles have boomed in Alaska thanks to 20 years of warm summers. The insects have chewed up 4 million acres of spruce trees.

Other effects could happen later this century, if warming continues:

1. Spread of Disease

Warmer temperatures along with associated floods and droughts are encouraging worldwide health threats by creating an environment where mosquitoes, ticks, mice and other disease-carrying creatures thrive. The World Health Organization (WHO) reports that outbreaks of new or resurgent diseases are on the rise and in more disparate countries than ever before, including tropical illnesses in once cold climates -- such as mosquitoes infecting Canadians with West Nile virus. While more than 150,000 people die from climate change-related sickness each year, everything from heat-related heart and respiratory problems to malaria are on the rise. Cases of allergies and asthma are also increasing. Global warming fosters increased smog -- which is linked to mounting instances of asthma attacks -- and also advances weed growth, a bane for allergy sufferers.

2. Warmer Waters and More Hurricanes

As the temperature of oceans rises, so will the probability of more frequent and stronger hurricanes. These hurricanes will cause great loss of life and resources.

3. Droughts and Heat Waves

Although some areas of Earth will become wetter due to global warming, other areas will suffer serious droughts and heat waves. Africa will receive the worst of it, with more severe droughts also expected in Europe. Water is already a dangerously rare commodity in Africa, and according to the Intergovernmental Panel on Climate Change, global warming will exacerbate the conditions and could lead to conflicts and war. The deadly heat wave that swept across Europe in 2003, killing an estimated 35,000 people, could be the harbinger of an intense heat trend that scientists began tracking in the early 1900s. Extreme heat waves are happening two to four times more often now, steadily rising over the last 50 to 100 years, and are projected to be 100 times more likely over the next 40 years. Experts suggest continued heat waves may mean future increases in wildfires, heat-related illness and a general rise in the planet's mean temperature.

4. Melting of Polar Ice Caps

The ice cap melting is a four-pronged danger. First, it will raise sea levels. There are 5,773,000 cubic miles of water in ice caps, glaciers, and permanent snow. According to the National Snow and Ice Data Center, if all glaciers melted today the seas would rise about 230 feet. Second, melting ice caps will throw the global ecosystem out of balance. The ice caps are fresh water, and when they melt they will desalinate the ocean, or in plain English - make it less salty. The desalinization of the Gulf current will "mess up" ocean currents, which regulate temperatures. The stream shutdown or irregularity would cool the area around Northeast America and Western Europe. Earth's hotter temperature doesn't necessarily mean the Miami lifestyle is moving to the Arctic, but it does mean rising sea levels. Hotter temperatures mean ice - glaciers, sea ice and polar ice sheets - is melting, increasing the amount of water in the world's seas and oceans.

Scientists are able to measure that melt water from Greenland's ice cap directly impacts people in the United States: The flow of the Colorado River has increased six fold. And scientists project that as the ice shelves on Greenland and Antarctica melt, sea levels could be more than 20 feet (6 meters) higher in 2100 than they are today. Such levels would submerge many of Indonesia's tropical islands and flood low-lying areas such as Miami, New York City's Lower Manhattan and Bangladesh. Tundra once covered with thick permafrost is melting with rising surface temperatures and is now coated with plant life. In the span of a century, glaciers in Montana's Glacier National Park have deteriorated from 150 to just 35. And the Himalayan glaciers that feed the Ganges River, which supplies drinking and irrigation water to 500 million people, are reportedly shrinking by 40 yards (37 meters) each year.

5. More Floods

Flooding represents one of the most dangerous hazards to human settlements and is one of the most potentially momentous impacts of global warming. As the climate changes, a warming of the seas creates 'thermal expansion'. This is where warm water begins to take up more space than cool water, making the sea's surface level increase. Thermal expansion has already raised the height of the oceans by 4 to 8 inches (10 to 20cm).

Steadily melting glacial ice also adds significantly to the elevation in water surface level, and many low-lying or coastal communities and facilities will be under threat of eradication should the sea levels continue to rise. An increase of just a single meter (3 ft) would submerge considerable sections of the U.S. eastern seaboard, while one sixth of Bangladesh could be lost permanently by a rise of 1.5 m (5 ft), to name just two examples. The relocation of power stations, refineries, hospitals, homes and so on would become an expensive priority. Also, warmer air can hold more water vapor, increasing the level of rainfall and bringing flooding to inland areas.

6. Destructive Storms

With ocean temperature being a key factor for hurricane formation, the consequences of global warming will inevitably include the increased generation of storms and hurricanes with greater power and frequency. The destructive power of hurricanes has increased by some 50% in the last 30 years, a figure that is closely connected with the rising temperature of the ocean. Warmer water leads to greater evaporation, which in turn helps to not just 'prime' the coalescence of hurricanes and cyclones but also to maintain their vigor once extant.

7. Fires and Wildfires

As the planet continues to warm, dry areas of land that are already susceptible to wildfires are likely to be ravaged by even more frequent and destructive episodes. In 2007, more than 3,000 fires brought destruction to Southeastern Europe thanks to a long summer that created arid and parched conditions – a situation that would become normal as a consequence of the greenhouse effect. What's more, the carbon dioxide and 'black carbon' (a very fine soot) released by these large-scale fires together with the deforestation they cause further compounds the problem of air pollution – as the gases that help to create the greenhouse effect are supplemented and less mature trees survive to draw CO_2 from the atmosphere.

8. Death by Smog

A powerful combination of vehicular fumes, ground-level ozone, airborne industrial pollution and the stagnant hot air associated with heat waves, smog represents an immediate and chronic health threat to those living in built-up urban areas. It exacerbates pre-existing health conditions that affect the respiratory system such as emphysema, bronchitis and asthma, and in general impedes the immune system's ability to fight against infection and disease.

9. Tsunamis

Although global warming does not directly influence the formation of tsunamis, they can be generated by events that are brought about by an amplification of the planet's temperature. One example is the melting of ice sheets. Being extremely heavy, massive glaciers apply a considerable amount of pressure to the Earth's surface underneath them. This anchorage decreases as the glaciers diminish, resulting in a 'freeing up' of tectonic masses that can lead to massive earthquakes and significant volcanic activity, both of which are capable of creating deadly tsunamis.

10. Cold Waves

A cold wave is characterized by a major plunge in temperature over a 24 hour period. It can be a devastating shock for crops and commerce, and also bring death and injury to humans and animals through accidents, hypothermia and starvation. Damage to pipelines and property can be costly, and, particularly if snowfall accompanies the cold wave, transport systems can grind to a halt, adversely affecting the distribution of food, water and medical supplies.

More than 150 people lost their lives during the 2009 to 2010 winter after record low temperatures and abundant snowfall caused disruption to much of Europe – which doesn't take into account the many thousands more excess winter deaths that were caused indirectly. It was the UK's coldest winter for three decades. It may seem illogical at first to attribute harsher cold weather to global warming, but a change in atmospheric patterns brought about by receding glacial ice can lead to the redirection of polar air currents and the sun's rays being absorbed by the larger areas of dark blue sea, while critical phenomena like the Gulf Stream can be affected by changing ocean temperatures as well.

11. Increased Volcanic Activity

As already noted, melting glaciations can usher in new, more frequent and more dangerous episodes of volcanic activity. The shifting pressures brought about by the lightening of the vast ice sheets allows the Earth's crust to 'bounce back' and can cause eruptions in unexpected places – like the one experienced during Iceland's Gjalp eruption, where magma reached the surface at an unusual intermediary point between two volcanoes. Potent or sustained volcanic activity can have an immense impact on human life even if the activity is centered away from dense population centers. It also has the potential to affect the planet's climate by injecting tons of gases and solids into the atmosphere that can remain there for weeks.

12. More Dangerous Thunderstorms

A consequence of the increased amounts of humid air generated by global warming is that more thunderstorms will be triggered. Research into the dynamic between climate change and thunderstorm power and frequency suggests that by the end of the century the occurrence of major thunderstorms could rise by over 100% in some places. Not only that, but this increase would generally occur during the existing storm season and not at times when such storms might provide beneficial rainfall to arid areas. Thunderstorms are also a common way of starting the devastating wildfires mentioned above.

13. Economic Consequences

The costs associated with climate change rise along with the temperatures. Severe storms and floods combined with agricultural losses cause billions of dollars in damages, and money is needed to treat and control the spread of disease. Extreme weather can create extreme financial setbacks.

Economic considerations reach into nearly every facet of our lives. Consumers face rising food and energy costs along with increased insurance premiums for health and home. Governments suffer the consequences of diminished tourism and industrial profits, soaring energy, food and water demands, disaster cleanup and border tensions.

14. Migration, Conflict and Wars

It is possible that future centuries could see increased friction between nations and ethnic groups as dwindling resources lead to migration and conflict. Countries and factions would seek to control precious, dwindling resources and provide safety and shelter for their own people – perhaps at the cost of others. Simultaneously, previously heavily populated places would become uninhabitable due to heat or other factors, displacing millions of people. These refugee hordes might be corralled into semi-permanent camps, or even suffer at the hands of unwelcoming native groups.

Even now, relocations are taking place. Mumbai's population is estimated to become swollen by a further 7 million people by the year 2050 as global warming renders villages and hamlets uninhabitable or unprofitable, either through flooding or drought. More land pollution would be an inevitable by-product of these changes in habitation and the availability of resources.

Declining amounts of quality food, water and land may be leading to an increase in global security threats, conflict and war. Scientists and military analysts alike are theorizing climate change and its consequences such as food and water instability pose threats for war and conflict, suggesting that violence and ecological crises are entangled. Countries suffering from water shortages and crop loss become vulnerable to security trouble, including regional instability, panic and aggression.

15. Death of Ocean Life

The world's oceans absorb roughly 30% of all anthropogenic carbon dioxide that seeps into the atmosphere, and so inevitably, as more fossil fuels are burned, ocean life will continue to suffer the negative consequences of global warming. One of the most critical changes brought about by global warming is the ongoing reduction of phytoplankton. These tiny plants are an integral food source for ocean life and are responsible for around half of the world's photosynthetic activity. Essentially, they are the foundations of the oceanic food chain, so a reduction in their numbers creates a knock-on effect that ripples up the entire food chain, particularly affecting the predators at the top.

Additionally, ocean acidification and warmer surface temperatures increase the dangers to many aquatic animals, particularly crustaceans, molluscs and coral reefs. Coral reefs are very sensitive to temperature changes, with many of them already observed to have 'bleached' and died.

16. Loss of Biodiversity and Animal Extinction

Loss of habitat for polar-ice edge communities such as polar bears is perhaps the most obvious consequence of having a warmer climate. Animals that are entirely dependent on cold environments will retreat to more northerly locations as the planet heats up – leading to encroachment upon other ecosystems and displacement of other animals from their natural habitat. A strong connection between oceanic warming, declines in reproduction and increases in mortality rates among seabirds, seals and sea lions has already been observed.

Acid rain has also been identified as having an adverse influence. One example of this is the death of large amounts of snails in areas prone to acidic precipitation. Birds dependent upon the snails as a calcium-rich food source and, without a suitable replacement for this loss to their diet, lay eggs with a much higher amount of defective shells. Humans also aren't immune to the threat. Desertification and rising sea levels threaten human habitats. And when plants and animals are lost to climate change, human food, fuel and income are lost as well.

17. Destruction of Ecosystems

Changing climatic conditions and dramatic increases in carbon dioxide will put our ecosystems to the test, threatening supplies of fresh water, clean air, fuel and energy resources, food, medicine and other matters we depend upon not just for our lifestyles but for our survival.

Evidence shows effects of climate change on physical and biological systems, which means no part of the world is spared from the impact of changes to land, water and life. Scientists are already observing the bleaching and death of coral reefs due to warming ocean waters, as well as the migration of vulnerable plants and animals to alternate geographic ranges due to rising air and water temperatures and melting ice sheets. Models based on varied temperature increases predict scenarios of devastating floods, drought, wildfires, ocean acidification and eventual collapse of functioning ecosystems worldwide, terrestrial and aquatic alike.

Forecasts of famine, war and death paint a dire picture of climate change on our planet. Scientists are researching the causes of these changes the vulnerability of Earth not to predict the end of days but rather to help us mitigate or reduce changes that may be caused by humans. If we know and understand the problems and take action through adaptation, the use of more energy-efficient and sustainable resources and the adoption of other green ways of living, we may be able to make some impact on the climate change process.

Effect of Global Warming in India

India has a vast coastal line and the rising sea levels caused by global warming will cause an ecological disaster. The Himalayan glaciers have started to melt and the average rate of retreat is almost twice (34 meters) per year as compared to the 1971 levels of 19 meters. The melting glaciers will cause temperatures and sea-levels to rise and there will be a cascading effect on the crops and the monsoons. Worse – whole islands are expected to vanish! In fact two have already gone under – two islands in the Sunderbans, an area which India shares with Bangladesh. A temperature in the group of islands has already gone up by one degree centigrade.

The east coast of India being affected more because the Bay of Bengal is landlocked from three sides and there is a huge delta of the rivers Brahmaputra and the Ganga. These rivers will carry the water from the melting Himalayan snows. However this does not mean that the western coastal regions are immune…just that the eastern coast is more vulnerable at this stage.

Preventive Steps to Reduce Global Warming

Burning fossil fuels such as natural gas, coal, oil and gasoline raises the level of carbon dioxide in the atmosphere, and carbon dioxide is a major contributor to the greenhouse effect and global warming. We can help to reduce the demand for fossil fuels, which in turn reduces global warming, by using energy more wisely. Here are simple actions that can help to reduce global warming.

Reduce, Reuse, Recycle

Reducing, reusing and recycling minimize the amount of waste people generate. The items people throw out all take energy to make; many of them are not biodegradable and may take centuries to break down. Reducing is the first and most effective of the three Rs.

Reduce

It means reducing your consumption or buying less. Designing items like plastic bottles in ways that use less material is another way to reduce consumption. Using steel cutlery instead of plastic utensils, buying used goods, mending clothes instead of buying new ones and consuming less electricity are all examples of ways one can reduce in own life.

Reuse

Rather than throwing out items like clothing or food jars, consumers can find new uses for them -- and thereby reduce their consumption of new resources. Composting, using jars to store beverages or leftover food, and trading or selling used DVDs rather than throwing them out are all examples of ways people can reuse. Reusing is the second most effective of the three Rs; like reducing, it avoids creating waste rather than trying to recycle it once it's already there.

Recycling

Recycling is the third of the three Rs. Recycling extracts valuable materials from items that might otherwise be considered trash and turns them into new products. Communities have a variety of recycling programs, such as curbside pickup of recyclables, drop-off centers, buy-back centers that pay you for valuable items and deposit-refund programs. Deposit-refund programs, which include a deposit as part of the product price, refund consumers when they recycle such items as soda cans and plastic bottles. As a consumer, one can also help recycling by purchasing products made from recycled material, such as toilet paper made from recycled pulp.

Besides these following things an also help in reducing global climate changes:

1) Lesser use of heat and air conditioning.
2) Use of energy efficient electrical appliances and other products.
3) Smart driving and more use of bicycles and public transport system.
4) Afforestation and reducing deforestation.
5) Encouraging others to use above mentioned practices.

BIOTECHNOLOGY: A SOLUTION PROVIDER FOR CLIMATE CHANGE

Since the industrial revolution, economic growth has been inextricably linked with accelerating negative environmental impact. The more mankind has produced the more the planet has been exploited. Biotechnology challenges this pattern and breaks the cycle of resource consumption by rethinking traditional industrial processes. By providing a range of options for competitive industrial performance, biotechnology could enhance economic growth, while at the same time saving water, energy, raw materials and reducing waste production. Biotechnology provides an essential toolbox of solutions in the task of mitigating the impacts of climate change. By utilizing biotechnologies greenhouse gas emissions can be reduced and cleaner and more sustainable energy resources can be incorporated.

Industrial or white biotechnology uses enzymes and micro-organisms to make biobased products in sectors such as chemicals, food and feed, detergents, paper and pulp, textiles and bioenergy (such as biofuels or biogas). In doing so, it uses renewable raw materials and is one of the most promising, innovative approaches towards lowering greenhouse gas emissions. The application of industrial biotechnology has been proven to make significant contributions towards mitigating the impacts of climate change in these and other sectors **(Fig 7.2)**. In addition to environmental benefits, biotechnology can improve industry's performance and product value and, as the technology develops and matures, white biotechnology will yield more and more viable solutions for our environment. These innovative solutions bring added benefits for both our climate and our economy.

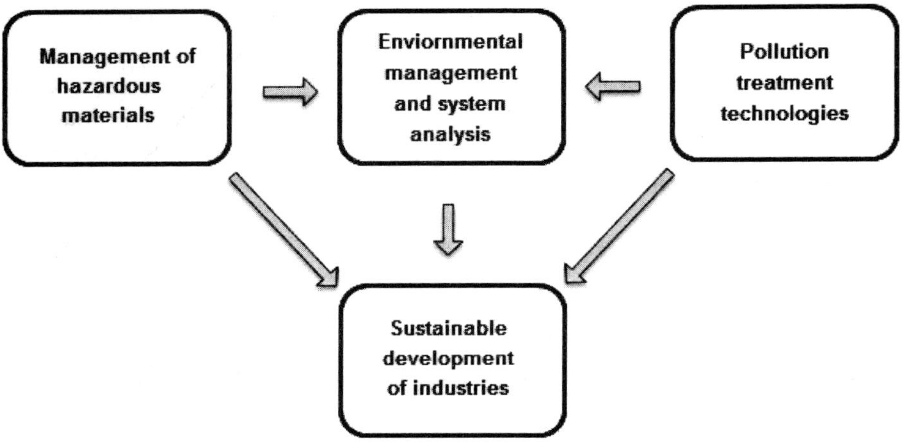

Fig. 7.2: Management of Global climate changes through biotechnology

Various points have been given below regarding role of biotechnology and industrial sustainability in producing clean industrial products:

Biotechnology and CO_2 Emissions

Fossil carbon represents the single most important raw material for energy generation and for chemicals manufacture, but its oxidation product, CO2, is an important greenhouse gas. Any means of reducing fossil carbon consumption, either by improving energy efficiency or by substituting alternative resources will directly result in lowered CO_2 production and thus reduce global warming.

Industrial Processes

Use of biotechnology has already resulted in energy reduction in industrial processes. There are many examples for the role of biotechnology in industries, like:

1) Ammonium acrylate, a key intermediate in the manufacture of acrylic polymers, is made by hydrolyzing acrylonitrile to acrylic acid and reacting this with ammonia. The reaction is energy-intensive and gives rise to by-products which are difficult to remove. A process, based on a bacterial enzyme which directly synthesizes ammonium

acrylate of the same quality under less energy-demanding conditions, has been operating for several years at full scale.

2) In paper making, treating cellulose fibers in the pulp using cellulase and hemicellulase enzymes allows water to drain more quickly from the wet pulp, thereby reducing processing time and energy used for drying.

Materials

Biomass, as it grows, consumes CO_2. Substances made from such renewable raw materials are therefore a zero net contributor to atmospheric greenhouse gases, unless fossil fuel is used in their manufacture. A wide range of chemicals and structural materials can be based on biological raw materials including biodegradable plastics, biopolymers and biopesticides, novel fibers and timbers. Plant-derived amides, esters and acetates are currently being used as plasticizers, blocking/slip agents and mould-release agents for synthetic polymers. Uses of biohydrocarbons linked to amines, alcohols, phosphates and sulphur ligands include fabric softeners, corrosion inhibitors, ink carriers, solvents, hair conditioners, and perfumes.

Clean Fuels

Bioethanol is a CO_2-neutral alternative liquid transportation fuel. As new technologies – including continuous fermentation, production from lignocellulosic (wood and agricultural crop) waste – and more efficient separation techniques are developed, the cost of bioethanol will compete with that of gasoline. Over a 20-year period, US ethanol production, based solely on lignocellulosic waste, could rise to 470 million tonnes a year, equal to present gasoline consumption in energy terms.

There is requirement of a shift from present petrochemical industry processes, which consume large quantities of energy under conditions of high temperature and pressure, to more energy-efficient biological processes, which use renewable resources such as biomass to produce useful substances under normal temperatures and pressures. For example, future processes will focus more on producing efficiently alternative fuels such as ethanol, which contribute less to global warming and are also likely to produce environmentally benign products, such as biodegradable plastics, which breaks down in natural settings after use. As a result, biotechnology should become an increasingly valuable tool for developing environmentally friendly products and processes and for preventing the Earth from warming.

ACID RAINS

"Acid Rain," or more precisely acid precipitation, is the word used to describe rainfall that has a pH level of less than 5.6. This form of air pollution is currently a subject of great controversy because of its worldwide environmental damages. For the last ten years, this phenomenon has brought destruction to thousands of lakes and streams in the United States, Canada, and parts of Europe. Acid rain is formed when oxides of nitrogen and sulfite combine with moisture in the atmosphere to make nitric and sulfuric acids. These acids can be carried away far from its origin. Acid rain is rain consisting of water droplets that are unusually acidic because of atmospheric pollution - most notably the excessive amounts of sulfur and nitrogen released by cars and industrial processes. Acid rain is also called acid deposition because this term includes other forms of acidic precipitation such as snow.

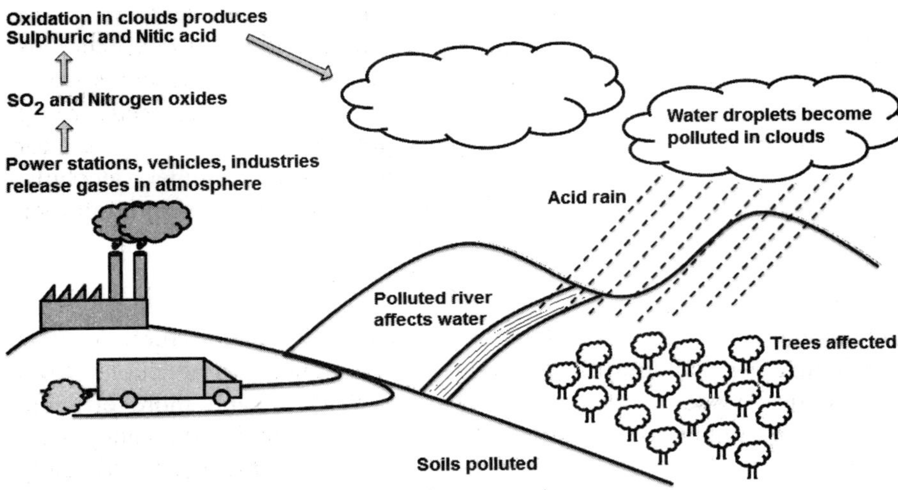

Fig. 7.3: Acid Rain formation due to SO_2 and oxides of nitrogen and its various effects

Acidic deposition occurs in two ways: wet and dry. Wet deposition is any form of precipitation that removes acids from the atmosphere and deposits them on the Earth's surface. Dry deposition polluting particles and gases stick to the ground via dust and smoke in the absence of precipitation. This form of deposition is dangerous however because precipitation can eventually wash pollutants into streams, lakes, and rivers. The two primary sources of acid rain are sulfur dioxide (SO_2), and oxides of nitrogen (NOx) **(Fig 7.3)**.

Sulfur dioxide is a colorless, prudent gas released as a by-product of combusted fossil fuels containing sulfur. A variety of industrial processes, such as the production of iron and steel, utility factories, and crude oil processing produce this gas. In iron and steel production, the smelting of metal sulfate ore, produces pure metal. This causes the release of sulfur dioxide. Metals such as zinc, nickel, and copper are commonly obtained by this process. Sulfur dioxide can also be emitted into the atmosphere by natural disasters or means. This ten percent of all sulfur dioxide emission comes from volcanoes, sea spray, plankton, and rotting vegetation. Overall,

69.4 percent of sulfur dioxide is produced by industrial combustion. Only 3.7 percent is caused by transportation.

The other chemical that is also chiefly responsible for the make-up of acid rain is nitrogen oxide. Oxide of nitrogen is a term used to describe any compound of nitrogen with any amount of oxygen atoms. Nitrogen monoxide and nitrogen dioxide are all oxides of nitrogen. These gases are by-products of firing processes of extreme high temperatures (automobiles, utility plants), and in chemical industries (fertilizer production). Natural processes such as bacterial action in soil, forest fires, volcanic action, and lightning make up five percent of nitrogen oxide emission. Transportation makes up 43 percent, and 32 percent belongs to industrial combustion. Nitrogen oxide is a dangerous gas by itself. This gas attacks the membranes of the respiratory organs and increases the likelihood of respiratory illness. It also contributes to ozone damage, and forms smog. Nitrogen oxide can spread far from the location it was originated by acid rain. Combustion of fuels produces sulphur dioxide and nitric oxides. They are converted into sulfuric acid and nitric acid. In the gas phase sulphur dioxide is oxidized by reaction with the hydroxyl radical via an intermolecular reaction:

$$\mathbf{SO_2 + OH \rightarrow HOSO_2} \quad \text{followed by}$$

$$\mathbf{HOSO_2 + O_2 \rightarrow H_2O + SO_3}$$

In the presence of water, sulfur trioxide (SO_3) is converted rapidly to sulfuric acid:

$$\mathbf{SO_3\,(g) + H_2O\,(l) \rightarrow H_2SO_4\,(aq)}$$

Nitrogen dioxide reacts with OH to form nitric acid:

$$\mathbf{NO_2 + OH \rightarrow HNO_3}$$

Today, acid deposition is present in the northeastern United States, southeastern Canada, and much of Europe including portions of Sweden, Norway, and Germany. In addition, parts of South Asia, South Africa, Sri Lanka, and Southern India are all in danger of being impacted by acid deposition in the future.

Effects of Acid Rain

Acid rain causes acidification of lakes and streams and contributes to the damage of trees at high elevations (for example, red spruce trees above 2,000 feet) and many sensitive forest soils. In addition, acid rain accelerates the decay of building materials and paints, including irreplaceable buildings, statues, and sculptures that are part of our nation's cultural heritage. Prior to falling to the earth, sulfur dioxide (SO_2) and nitrogen oxide (NOx) gases and their particulate matter derivatives—sulfates and nitrates—contribute to visibility degradation and harm public health. Acid rain can affect the earth in many different ways. It can affect soil, buildings, lakes, rivers, and human health in various ways **(Fig 7.3)**.

Soil

Acid rain can damage soil by destroying many vital substances and washing away the nutrients. Soils naturally contain small amounts of poisonous minerals such as mercury and aluminum. Normally these minerals do not cause serious problems, but when acid rain falls on the ground and the acidity of the soil increases, chemical reactions occur allowing the poisonous minerals to be taken up by the plant roots. The trees and plants are then damaged and any animals eating them will absorb the poisons, which will stay in their bodies.

Surface Waters and Aquatic Animals

The ecological effects of acid rain are most clearly seen in the aquatic, or water, environments, such as streams, lakes, and marshes. Acid rain flows into streams, lakes, and marshes after falling on forests, fields, buildings, and roads. Acid rain also falls directly on aquatic habitats. Most lakes and streams have a pH between 6 and 8, although some lakes are naturally acidic even without the effects of acid rain. Acid rain primarily affects sensitive bodies of water, which are located in watersheds whose soils have a limited ability to neutralize acidic compounds (called "buffering capacity"). Lakes and streams become acidic (i.e., the pH value goes down) when the water itself and its surrounding soil cannot buffer the acid rain enough to neutralize it. In areas where buffering capacity is low, acid rain releases aluminum from soils into lakes and streams; aluminum is highly toxic to many species of aquatic organisms.

Acid rain causes a cascade of effects that harm or kill individual fish, reduce fish population numbers, completely eliminate fish species from a water body, and decrease biodiversity. As acid rain flows through soils in a watershed, aluminum is released from soils into the lakes and streams located in that watershed. So, as pH in a lake or stream decreases, aluminum levels increase. Both low pH and increased aluminum levels are directly toxic to fish. In addition, low pH and increased aluminum levels cause chronic stress that may not kill individual fish, but leads to lower body weight and smaller size and makes fish less able to compete for food and habitat. Some types of plants and animals are able to tolerate acidic waters. Others, however, are acid-sensitive and will be lost as the pH declines. Generally, the young of most species are more sensitive to environmental conditions than adults. At pH 5, most fish eggs cannot hatch. At lower pH levels, some adult fish die. Some acid lakes have no fish. Together, biological organisms and the environment in which they live are called an ecosystem. The plants and animals living within an ecosystem are highly interdependent. For example, frogs may tolerate relatively high levels of acidity, but if they eat insects like the mayfly, they may be affected because part of their food supply may disappear. Because of the connections between the many fish, plants, and other organisms living in an aquatic ecosystem, changes in pH or aluminum levels affect biodiversity as well. Thus, as lakes and streams become more acidic, the numbers and types of fish and other aquatic plants and animals that live in these waters decrease.

The impact of nitrogen on surface waters is also critical. Nitrogen plays a significant role in episodic acidification and new research recognizes the importance of nitrogen in long-term chronic acidification as well. Furthermore, the adverse impact of atmospheric nitrogen deposition on estuaries and near-coastal water bodies is significant. Scientists estimate that 10 to 45 percent of the nitrogen produced by various human activities that reaches estuaries and coastal ecosystems is transported and deposited via the atmosphere. Nitrogen is an important factor in causing eutrophication (oxygen depletion) of water bodies. The symptoms of eutrophication include blooms of algae (both toxic and non-toxic), declines in the health of fish and shellfish, loss of sea grass beds and coral reefs, and ecological changes in food webs. These ecological changes impact human populations by changing the availability of seafood and creating a risk of consuming contaminated fish or shellfish, reducing our ability to use and enjoy our coastal ecosystems, and causing economic impact on people who rely on healthy coastal ecosystems, such as fishermen and those who cater to tourists.

Forests

Over the years, scientists, foresters, and others have noted a slowed growth of some forests. Leaves and needles turn brown and fall off when they should be green and healthy. In extreme cases, individual trees or entire areas of the forest simply die off without an obvious reason.

After much analysis, researchers now know that acid rain causes slower growth, injury, or death of forests. Acid rain has been implicated in forest and soil degradation in many areas of the world; particularly high elevation forests of the Appalachian Mountains from Maine to Georgia that include areas such as the Shenandoah and Great Smoky Mountain National Parks. Of course, acid rain is not the only cause of such conditions. Other factors contribute to the overall stress of these areas, including air pollutants, insects, disease, drought, or very cold weather. In most cases, in fact, the impacts of acid rain on trees are due to the combined effects of acid rain and these other environmental stressors. After many years of collecting information on the chemistry and biology of forests, researchers are beginning to understand how acid rain works on the forest soil, trees, and other plants.

Acid rain does not usually kill trees directly. Instead, it is more likely to weaken trees by damaging their leaves, limiting the nutrients available to them, or exposing them to toxic substances slowly released from the soil. Quite often, injury or death of trees is a result of these effects of acid rain in combination with one or more additional threats.

Scientists know that acidic water dissolves the nutrients and helpful minerals in the soil and then washes them away before trees and other plants can use them to grow. At the same time, acid rain causes the release of substances that are toxic to trees and plants, such as aluminum, into the soil. Scientists believe that this combination of loss of soil nutrients and increase of toxic aluminum may be one way that acid rain harms trees. Such substances also wash away in the runoff and are carried into streams, rivers, and lakes. More of these substances are released from the soil when the rainfall is more acidic.

Acid rain can harm other plants in the same way it harms trees. Although damaged by other air pollutants such as ground level ozone, food crops are not usually seriously affected because farmers frequently add fertilizers to the soil to replace nutrients that have washed away. They may also add crushed limestone to the soil. Limestone is an alkaline material and increases the ability of the soil to act as a buffer against acidity.

Automotive Coating

Over the past two decades, there have been numerous reports of damage to automotive paints and other coatings. The reported damage typically occurs on horizontal surfaces and appears as irregularly shaped, permanently etched areas. The general consensus within the auto industry is that some form of environmental fallout causes the damage. "Environmental fallout"—a term widely used in the auto and coatings industries—refers to damage caused by air pollution (e.g., acid rain), decaying insects, bird droppings, pollen, and tree sap. The results of laboratory experiments and at least one field study have demonstrated that acid rain can scar automotive coatings. Furthermore, chemical analyses of the damaged areas of some exposed test panels indicate elevated levels of sulfate, implicating acid rain. All forms of acid rain, including dry deposition, especially when dry acidic deposition is mixed with dew or rain, may damage automotive coatings. However, it has been difficult to quantify the specific contribution of acid rain to paint finish damage relative to damage caused by other forms of environmental fallout, by the improper application of paint or by deficient paint formulations. According to coating experts, trained specialists can differentiate between the various forms of damage, but the best way of determining the cause of chemically induced damage is to conduct a detailed, chemical analysis of the damaged area.

Buildings and Monuments

When sulphur pollutants fall on to buildings made from limestone and sandstone they react with minerals in the stone to form a powdery substance that can be washed away by rain. Famous buildings like the Statue of Liberty in New York, the Taj Mahal in India and St. Paul's Cathedral in London have all been damaged by this sort of air pollution.

Acid rain can also damage stained glass windows in churches, railway lines and steel bridges. The acid rain slowly eats away them all. Building materials crumble away, metals are corroded, the color of paint is spoiled, leather is weakened and crusts form on the surface of glass.

Human Health

Acid rain looks, feels, and tastes just like clean rain. The harm to people from acid rain is not direct. Walking in acid rain, or even swimming in an acid lake, is no more dangerous than walking or swimming in clean water. However, the pollutants that cause acid rain—sulfur dioxide (SO_2) and nitrogen oxides (NO_x)—do damage human health. These gases interact in the atmosphere to form fine sulfate and nitrate particles that can be transported long distances by winds and inhaled deep into people's lungs. Fine particles can also penetrate indoors. Many scientific studies have identified a relationship between elevated levels of fine particles and increased illness and premature death from heart and lung disorders, such as asthma and bronchitis. When we breathe in air pollution, the very fine particulates can easily enter our lungs, where they can cause breathing problems, and over time even lead to cancer. Drinking water is contaminated with chemicals released by acid rain. Aluminum and lead in water can be poisonous at high levels.

Acid Rain Prevention Methods

There are several ways to reduce acid rain - more properly called acid deposition—ranging from societal changes to individual action.

Clean up Smokestacks and Exhaust Pipes

Almost all of the electricity that powers modern life comes from burning fossil fuels such as coal, natural gas, and oil. Acid deposition is caused by two pollutants that are released into the atmosphere when fossil fuels are burned: sulfur dioxide (SO_2) and nitrogen oxides (NO_x). Coal accounts for most SO_2 emissions and a large portion of NO_x emissions. Sulfur is present in coal as an impurity, and it reacts with air when the coal is burned to form SO_2. In contrast, NO_x is formed when any fossil fuel is burned. There are several options for reducing SO_2 emissions, including using coal containing less sulfur, washing the coal, and using devices called "scrubbers" to chemically remove the SO_2 from the gases leaving the smokestack (explained in chapter 3). Power plants can also switch fuels—for example, burning natural gas creates much less SO_2 than burning coal. Certain approaches will also have the additional benefit of reducing other pollutants such as mercury and carbon dioxide (CO_2). Understanding these "co-benefits" has become important in seeking cost-effective air pollution reduction strategies. Finally, power plants can use technologies that do not burn fossil fuels. Each of these options, however, has its own costs and benefits; there is no single universal solution. Similar to scrubbers on power plants, catalytic converters reduce NO_x emissions from cars.

Use Alternative Energy Sources

There are other sources of electricity besides fossil fuels. They include nuclear power, hydropower, wind energy, geothermal energy, and solar energy. There are also alternative energies, such as natural gas, batteries, and fuel cells, available to power automobiles.

All sources of energy have environmental costs as well as benefits. Some types of energy are more expensive to produce than others, which mean that not all people can afford all of them. Nuclear power, hydropower, and coal are the cheapest forms of energy today, but advancements in technologies and regulatory developments may change this in the future. All of these factors must be weighed when deciding which energy source to use today and which to invest in for tomorrow.

Restore a Damaged Environment

Acid deposition penetrates deeply into the fabric of an ecosystem, changing the chemistry of the soil and streams and narrowing—sometimes to nothing—the space where certain plants and animals can survive. Because there are so many changes, it takes many years for ecosystems to recover from acid deposition, even after emissions are reduced and the rain pH is restored to normal. For example, while visibility might improve within days, and small or episodic chemical changes in streams improve within months, chronically acidified lakes, streams, forests, and soils can take years to decades, or even centuries (in the case of soils) to heal.

However, there are some things that people can do to bring back lakes and streams more quickly. Limestone or lime (a naturally occurring basic compound) can be added to acidic lakes to "cancel out" the acidity. This process, called liming, has been used extensively in Norway and Sweden. Liming tends to be expensive, has to be done repeatedly to keep the water from returning to its acidic condition, and is considered a short-term remedy in only specific areas, rather than an effort to reduce or prevent pollution. Furthermore, it does not solve the broader problems of changes in soil chemistry and forest health in the watershed, and it does nothing to address visibility reductions, materials damage, and risk to human health. However, liming does often permit fish to remain in a lake, allowing the native population to survive in place until emissions reductions reduce the amount of acid deposition in the area.

Individual Actions

It may seem like there is not much that one individual can do to stop acid deposition. However, like many environmental problems, acid deposition is caused by the cumulative actions of millions of individual people. Therefore, each individual can also reduce their contribution to the problem and become part of the solution. Individuals can contribute directly by conserving energy, since energy production causes the largest portion of the acid deposition problem. For example, one can:

1) Turn off lights, computers, and other appliances when not using them.
2) Use energy-efficient appliances: lighting, air conditioners, heaters, refrigerators, washing machines, etc.
3) Only use electric appliances when one need them.
4) Carpool, use public transportation, or better yet, walk or bicycle whenever possible
5) Buy vehicles with low NO_x emissions, and proper maintenance vehicle.

OZONE DEPLETION

The ozone layer is responsible for absorbing harmful ultraviolet rays, and preventing them from entering the Earth's atmosphere. However, various factors have led to the depletion and damage of this protective layer. Ozone is a colorless gas found in the upper atmosphere of the Earth. It is formed when oxygen molecules absorb ultraviolet photons, and undergo a chemical reaction known as photo dissociation or photolysis.

Three forms (or allotropes) of oxygen are involved in the ozone-oxygen cycle: oxygen atoms (O or atomic oxygen), oxygen gas (O_2 or diatomic oxygen), and ozone gas (O_3 or triatomic oxygen). Ozone is formed in the stratosphere when oxygen molecules photo dissociate after absorbing an ultraviolet photon whose wavelength is shorter than 240 nm. This converts a single O_2 into two atomic oxygen radicals. The atomic oxygen radicals then combine with separate O_2 molecules to create two O_3 molecules. These ozone molecules absorb UV light between 310 and 200 nm, following which ozone splits into a molecule of O_2 and an oxygen atom. The oxygen atom then joins up with an oxygen molecule to regenerate ozone. This is a continuing process that terminates when an oxygen atom "recombines" with an ozone molecule to make two O_2 molecules.

$$\mathbf{2\ O_3 \rightarrow 3O_2}$$

Ozone as a Natural Sun Block

The electromagnetic radiation released from sun comprises ultraviolet radiation that is potentially injurious to most living things as it can damage DNA. The ozone layer screens out the sun's injurious ultraviolet radiation. Even 1% decreases in the amount of ozone in the upper stratosphere causes a measurable raise in the ultraviolet radiation which reaches the earth surface **(Fig 7.4)**. When there was no ozone at all, the quantity of ultraviolet radiation reaching us would be catastrophically very high. All living beings would suffer radiation burns, except they were underground, or in the sea.

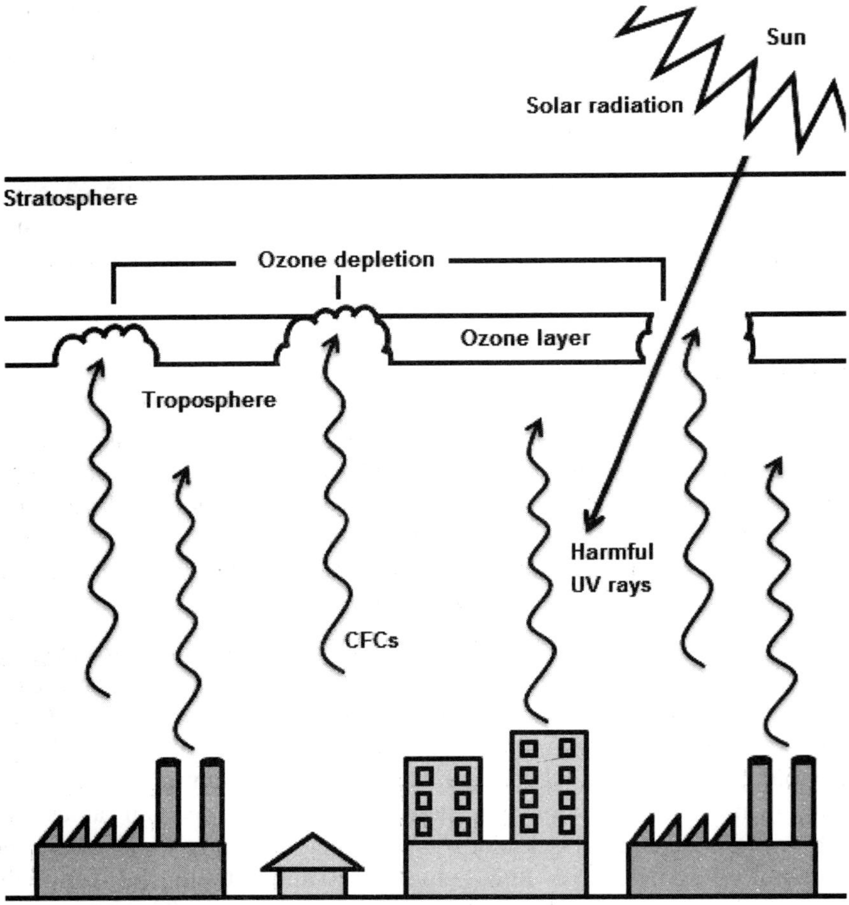

Fig. 7.4: Ozone layer depletion and penetration of UV rays

In stratosphere, small quantity of ozone is continuously being made by the action of sunlight on oxygen. At same time, ozone is being broken down by natural procedures. The net quantity of ozone generally stays constant as its formation and destruction takes place at about similar rate. Though unfortunately human action has recently changed that natural balance. Some manufactured substances like chlorofluoro carbons and hydrochloroflurocarbons can destroy stratosphere ozone much quicker than it is created.

Ultraviolet radiations (UVR), are high energy electromagnetic waves emitted from the Sun. UV radiation includes UV-A, the least dangerous form of UV radiation, UV-B, and UV-C, which is the most dangerous. UV-C is unable to reach the Earth's surface due to stratospheric ozone's ability to absorb it. The real threat comes from UV-B, which can enter the Earth's atmosphere, and has adverse effects. Ozone layer depletion first captured the attention of the whole world in the latter half of 1970, and since then, a lot of research has been done to find its possible effects and causes. Various studies have also been undertaken to find out a possible solution.

Causes of Ozone Depletion

Ozone can be destroyed by a number of free radical catalysts, the most important of which are the hydroxyl radical (OH·), the nitric oxide radical (NO·), atomic chlorine ion (Cl·) and atomic bromine ion (Br·).All of these have both natural and man-made sources; at the present time, most of the OH· and NO· in the stratosphere is of natural origin, but human activity has dramatically increased the levels of chlorine and bromine. These elements are found in certain stable organic compounds, especially chlorofluorocarbons (CFCs), which may find their way to the stratosphere without being destroyed in the troposphere due to their low reactivity. Once in the stratosphere, the Cl and Br atoms are liberated from the parent compounds by the action of ultraviolet light, e.g.

$$\mathbf{CFCl_3 + electromagnetic\ radiation \rightarrow CFCl_2 + Cl}$$

The Cl and Br atoms can then destroy ozone molecules through a variety of catalytic cycles. In the simplest example of such a cycle, a chlorine atom reacts with an ozone molecule, taking an oxygen atom with it (forming ClO) and leaving a normal oxygen molecule. The chlorine monoxide (i.e., the ClO) can react with a second molecule of ozone (i.e., O_3) to yield another chlorine atom and two molecules of oxygen. The chemical shorthand for these gas-phase reactions is **(Fig 7.5)**:

The chlorine atom changes an ozone molecule to ordinary oxygen

$$Cl + O_3 \rightarrow ClO + O_2$$

The ClO from the previous reaction destroys a second ozone molecule and recreates the original chlorine atom, which can repeat the first reaction and continue to destroy ozone.

$$ClO + O_3 \rightarrow Cl + 2O_2$$

The production and emission of chlorofluorocarbons (CFCs), is the leading cause of ozone layer depletion. CFC's accounts for almost 80% of the total depletion of ozone. Other ozone-depleting substances (ODS), include hydrochloroflurocarbons (HCFCs), and volatile organic compounds (VOCs). These are often found in vehicle emissions, byproducts of industrial processes, refrigerants, and aerosols. ODS are relatively stable in the lower atmosphere of the Earth, but in the stratosphere, they are exposed to ultraviolet radiation and thus, they break down to release a free chlorine atom.

This free chlorine atom reacts with an ozone molecule (O_3), and forms chlorine monoxide (ClO), and a molecule of oxygen. Now, ClO reacts with an ozone molecule to form a chlorine atom, and two molecules of oxygen. The free chlorine molecule again reacts with ozone to form chlorine monoxide. The process continues, and this results in the depletion of the ozone layer.

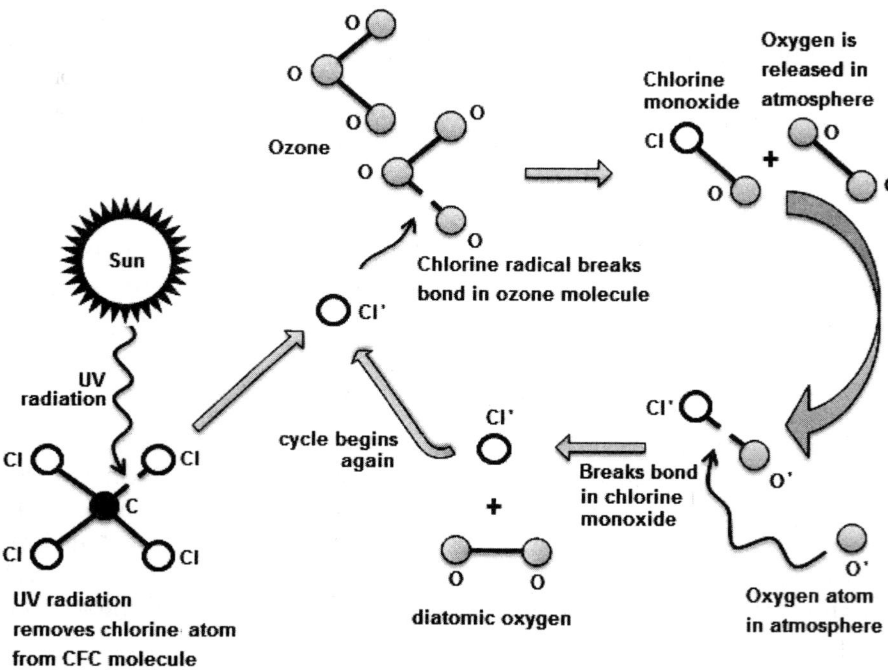

Fig. 7.5: Chemical reactions leading to breakdown of ozone molecules

CFCs and Related Compounds

Chlorofluorocarbons (CFCs) and other halogenated ozone depleting substances (ODS) are mainly responsible for man-made chemical ozone depletion. The total amount of effective halogens (chlorine and bromine) in the stratosphere can be calculated and are known as the equivalent effective stratospheric chlorine (EESC).

CFCs were invented by Thomas Midgley, Jr. in the 1920s. They were used in air conditioning and cooling units, as aerosol spray propellants prior to the 1970s, and in the cleaning processes of delicate electronic equipment. They also occur as by-products of some chemical processes. No significant natural sources have ever been identified for these compounds—their presence in the atmosphere is due almost entirely to human manufacture. As mentioned above, when such ozone-depleting chemicals reach the stratosphere, they are dissociated by ultraviolet light to release chlorine atoms. The chlorine atoms act as a catalyst, and each can break down tens of thousands of ozone molecules before being removed from the stratosphere. Given the longevity of CFC molecules, recovery times are measured in decades. It is calculated that a CFC molecule takes an average of about five to seven years to go from the ground level up to the upper atmosphere, and it can stay there for about a century, destroying up to one hundred thousand ozone molecules during that time.

Ozone Hole

Each spring in the stratosphere over Antarctica (spring in the southern hemisphere is from September through November.), atmospheric ozone is rapidly destroyed by chemical processes. As winter arrives, a vortex of winds develops around the pole and isolates the polar stratosphere. When temperatures drop below -78°C (-109°F), thin clouds form of ice, nitric acid, and sulphuric acid mixtures. Chemical reactions on the surfaces of ice crystals in the clouds release active forms of CFCs. Ozone depletion begins, and the ozone "hole" appears. Over the course of two to three months, approximately 50% of the total column amount of ozone in the atmosphere disappears. At some levels, the losses approach 90%. This has come to be called the Antarctic ozone hole. In spring, temperatures begin to rise, the ice evaporates, and the ozone layer starts to recover.

Possible Effects of Ozone Depletion

As ozone depletes in the stratosphere, it forms a 'hole' in the layer. This hole enables harmful ultraviolet rays to enter the Earth's atmosphere. Ultraviolet rays of the Sun are associated with a number of health-related and environmental issues. Let us take a look at how ozone depletion affects different life forms.

Impact on Humans

a) Skin cancer: Exposure to ultraviolet rays poses an increased risk of developing several types of skin cancers, including malignant melanoma, basal and squamous cell carcinoma.

b) Eye damage: Direct exposure to UV radiations can result in photokeratitis (snow blindness), and cataracts.

c) Immune system damage: Effects of UV rays include impairment of the immune system. Increased exposure to UV rays weakens the response of the immune system.

d) Accelerated aging of skin: Constant exposure to UV radiation can cause photo allergy, which results in the outbreak of rash in fair-skinned people.

e) Other effects: Ozone chemicals can cause difficulty in breathing, chest pain, throat irritation, and hamper lung functioning.

Effects on Amphibians
Ozone depletion is listed as one of the causes for the declining numbers of amphibian species. Ozone depletion affects many species of amphibians at every stage of their life cycle. Some of the effects are mentioned below.

a) Hampers growth and development in larvae.

b) Changes behavior and habits.

c) Causes deformities in some species.

d) Decreased immunity- Some species have become more vulnerable to diseases and death.

e) Retinal damage and blindness in some species.

Effects on Marine Ecosystems
In particular, plankton (phytoplankton and bacterio-plankton) are threatened by increased UV radiation. Marine phytoplankton plays a fundamental role in both the food chain as well as the oceanic carbon cycle. Plankton plays an important role in converting atmospheric carbon dioxide into oxygen. Ultraviolet rays can influence the survival rates of these microscopic organisms, by affecting their orientation and mobility. This eventually disturbs and affects the entire ecosystem.

Impact on Plants
In some species of plants, UV radiation can alter the time of flowering, as well as the number of flowers. Plant growth can be directly affected by UV-B radiation. Despite mechanisms to reduce or repair these effects, physiological and developmental processes of plants are affected.

Another observation is an increase in the ozone present in the lower atmosphere due to the decrease in the ozone in the stratosphere. Ozone present in the lower atmosphere is mainly regarded as a pollutant and a greenhouse gas that can contribute to global warming and climate change. However, studies have pointed out that the lifespan of lower atmospheric ozone is quite less, compared to stratospheric ozone. At the same time, increase in the level of ozone in the lower atmosphere can enhance the ability of sunlight to synthesize vitamin D, which can be regarded as an important beneficial effect of ozone layer depletion.

Growing concern for ozone depletion led to the adoption of the Montreal Protocol in 1987, in order to reduce and control industrial emission of chlorofluorocarbons (CFCs). Such international agreements have succeeded to a great extent in reducing the emission of these compounds. However, more cooperation and understanding among all the countries of the world is required to mitigate the problem.

Ozone Depletion and Global Warming

Ozone depletion gets worse when the stratosphere is very cold. It has been cold in the stratosphere in the last few years, so there has been particularly bad ozone depletion. Global temperatures during January - August 2011 were the third coldest on record in the lower stratosphere. In past few years ozone depletion has reached some of the most severe levels ever recorded over the Northern Hemisphere. Ozone absorbs the harmful solar UV radiation which results in heating of surrounding atmosphere. Ozone depletion leads to less ozone resulting in less UV radiation getting absorbed and less heating of atmosphere. Cooling of stratosphere leads to formation of polar stratospheric clouds which leads to further ozone destruction. For ozone levels to recover the atmosphere need to warm again. The ozone layer is not expected to begin recovery until around 2020 at the earliest.

Global Warming could supercharge ozone depletion. Ozone depletion gets worse when the stratosphere (where the ozone layer is) becomes colder. A significant portion of stratosphere cooling is also because of greenhouse gases like carbon dioxide and methane. Because climate change traps heat in the troposphere, less heat will reach the stratosphere which will make it colder. Greenhouse gases warm the troposphere and make the stratosphere colder. The global warming could make ozone depletion much worse right when it is supposed to begin its recovery during this century. Both ozone depletion and climate change have harmful effects on plants and animals. In every ecosystem, plants and animals are linked.

Ways to Protect the Ozone Layer

1) Minimize high altitude aircraft flights (oxygen reduction and water vapor deposition)
2) Minimize rocket flights (water vapor deposition)
3) Encourage growth of plants that produce oxygen, discourage deforestation
4) Decrease / control releases of high temperature steam / moisture to the atmosphere
5) Eliminate production and release of known ozone depleting chemicals (such as CFCs and HCFCs) where remotely possible. Subsidize production of safer alternatives where possible.
6) Establish controls to assure that new compounds to be used in high volume are surveyed for effect on ozone.

RADIOACTIVE WASTE MANAGEMENT

Radioactive wastes are wastes that contain radioactive material. Radioactive wastes are usually by-products of nuclear power generation and other applications of nuclear fission or nuclear technology, such as research and medicine. Radioactive waste is hazardous to most forms of life and the environment, and is regulated by government agencies in order to protect human health and the environment. Radioactivity diminishes over time, so waste is typically isolated and stored for a period of time until it no longer poses a hazard. The period of time waste must be stored depends on the type of waste. Radioactive wastes comprise a variety of materials requiring different types of management to protect people and the environment. The various radioactive isotopes have half-lives ranging from fractions of a second to minutes, hours or days, through to billions of years. Radioactivity decreases with time as these isotopes decay into stable, non-radioactive ones.

The rate of decay of an isotope is inversely proportional to its half-life; a short half life means that it decays rapidly. Hence, for each kind of radiation, the higher the intensity of radioactivity in a given amount of material, the shorter the half lives involved. Radioactivity arises naturally from the decay of particular forms of some elements, called isotopes. Some isotopes are radioactive, most are not, though here the focus is on the former. There are three kinds of radiation to consider: alpha, beta and gamma. A fourth kind, neutron radiation, generally only occurs inside a nuclear reactor. Different types of radiation require different forms of protection.

Alpha radiation cannot penetrate the skin and can be blocked out by a sheet of paper, but is dangerous in the lung. Beta radiation can penetrate into the body surface but can be blocked out by a sheet of aluminum foil. Gamma radiation can go deeply into the body and requires several centimeters of lead or concrete, or a meter or so of water, to block it.

All of these kinds of radiation are, at low levels, naturally part of our environment, where we are all naturally exposed to them at low levels. Any or all of them may be present in any classification of radioactive waste.

Types of Radioactive Waste

Low-level Waste

It is generated from hospitals, laboratories and industry, as well as the nuclear fuel cycle. It comprises paper, rags, tools, clothing, and filters etc. which contain small amounts of mostly short-lived radioactivity. It is not dangerous to handle, but must be disposed of more carefully than normal garbage. Usually it is buried in shallow landfill sites. To reduce its volume, it is often compacted or incinerated (in a closed container) before disposal.

Worldwide it comprises 90% of the volume but only 1% of the radioactivity of all radioactive waste.

Intermediate-level Waste

It contains higher amounts of radioactivity and may require special shielding. It typically comprises resins, chemical sludges and reactor components, as well as contaminated materials from reactor decommissioning. Worldwide it makes up 7% of the volume and has 4% of the radioactivity of all radioactive waste. It may be solidified in concrete or bitumen for disposal. Generally short-lived waste (mainly from reactors) is buried, but long-lived waste (from reprocessing nuclear fuel) is disposed of deep underground.

High-level Waste

It may be the used fuel itself, or the principal waste separated from reprocessing this. While only 3% of the volume of all radioactive waste, it holds 95% of the radioactivity. It contains the highly-radioactive fission products and some heavy elements with long-lived radioactivity. It generates a considerable amount of heat and requires cooling, as well as special shielding during handling and transport. If the used fuel is reprocessed, the separated waste is vitrified by incorporating it into borosilicate (Pyrex) glass which is sealed inside stainless steel canisters for eventual disposal deep underground.

On the other hand, if used reactor fuel is not reprocessed, all the highly-radioactive isotopes remain in it, and so the whole fuel assemblies are treated as high-level waste. This used fuel takes up about nine times the volume of equivalent vitrified high-level waste which is separated in reprocessing. Used fuel treated as waste must be encapsulated ready for disposal.

Both high-level waste and used fuel are very radioactive and people handling them must be shielded from their radiation. Such materials are shipped in special containers which shield the radiation and which will not rupture in an accident.

Whether used fuel is reprocessed or not, the volume of high-level waste is modest, - about 3 cubic meters per year of vitrified waste, or 25-30 tonnes of used fuel for a typical large nuclear reactor. The relatively small amount involved allows it to be effectively and economically isolated.

Radioactive Materials in the Natural Environment

Naturally-occurring radioactive materials are widespread throughout the environment, although concentrations are very low and they are not normally harmful. However, human activity may concentrate these so that they need careful handling- e.g. in coal ash and gas well residues.

Soil naturally contains a variety of radioactive materials - uranium, thorium, radium and the radioactive gas radon which is continually escaping to the atmosphere. Many parts of the Earth's crust are more radioactive than the low-level waste described above. Radiation is not something which arises just from using uranium to produce electricity, although the mining and milling of uranium and some other ores brings these radioactive materials into closer contact with people, and in the case of radon and its daughter products, speeds up their release to the atmosphere.

Radioactive Waste Handling

The general principles employed in the management of radioactive wastes are **(Fig 7.6)**:

1) Concentrate-and-contain
2) Dilute-and-disperse
3) Delay-and-decay
4) Reprocessing

If the used fuel is later reprocessed, it is dissolved and separated chemically into uranium, plutonium and high-level waste solutions. About 97% of the used fuel can be recycled leaving only 3% as high-level waste. The recyclable portion is mostly uranium depleted to less than 1% U-235, with some plutonium, which is most valuable.

Immobilizing Separated High-level Waste

Solidification processes have been developed in several countries over the past fifty years. Liquid high-level wastes are evaporated to solids, mixed with glass-forming materials, melted and poured into robust stainless steel canisters which are then sealed by welding. Long-term storage of radioactive waste requires the stabilization of the waste into a form which will neither react nor degrade for extended periods of time. One way to do this is through vitrification. Currently at Sellafield the high-level waste (PUREX first cycle raffinate) is mixed with sugar and then calcined. Calcination involves passing the waste through a heated, rotating tube. The purposes of calcination are to evaporate the water from the waste, and de-nitrate the fission products to assist the stability of the glass produced. This is termed as vitrification.

Because of the different characteristics of solids, liquids, and gases, each must be processed differently. The waste must also be processed in such a manner as to minimize the risk of exposure to the public **(Fig 7.6)**.

Liquids are processed to remove the radioactive impurities. These processes might include:

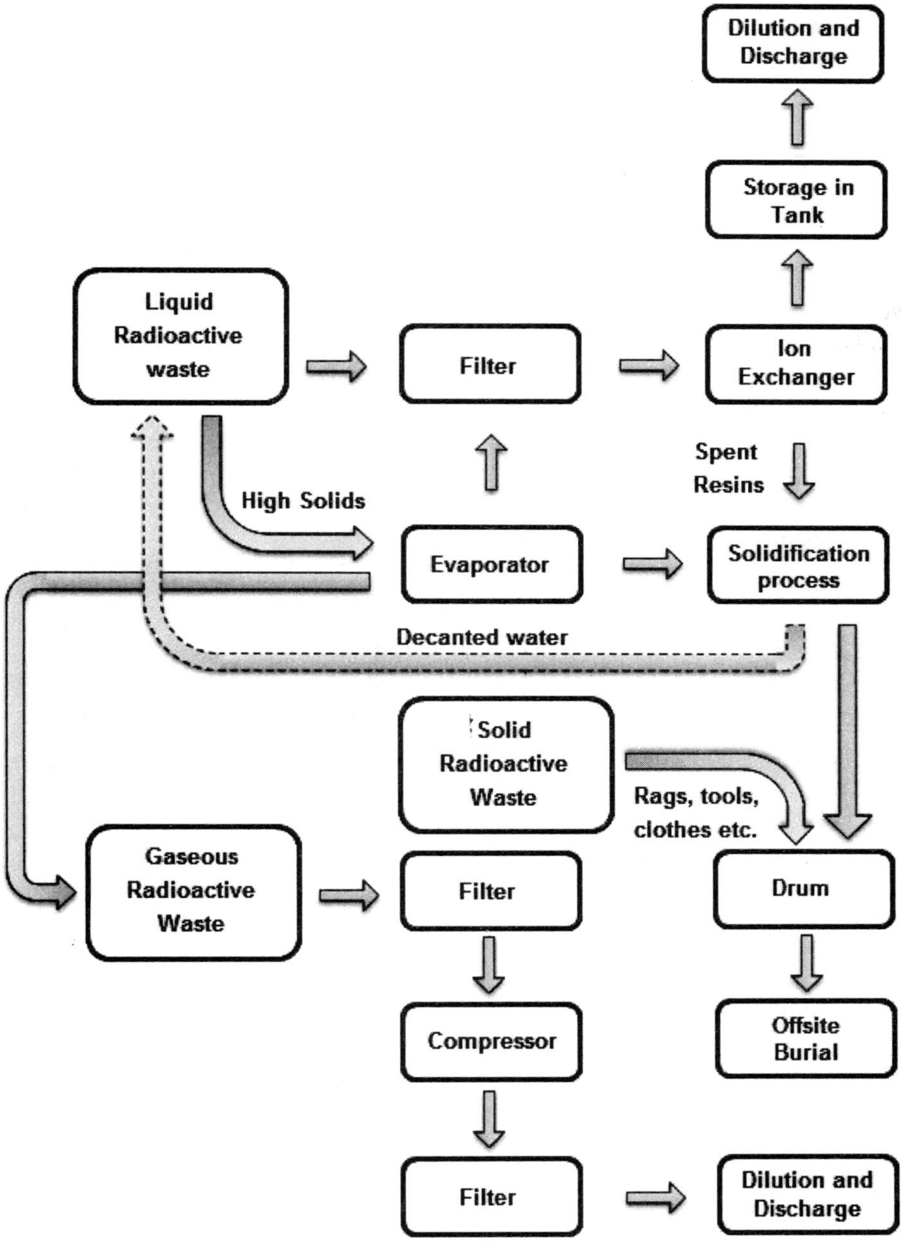

Fig. 7.6: Radioactive waste handling system

(i) Filtering,

(ii) Routing through demineralizer,

(iii) Boiling off the water (evaporation) and leaving the solid impurities (which are then processed as solid radioactive waste), and/or

(iv) Storing the liquid for a time period to allow the radioactive material to decay.

After processing, the water will be sampled. If samples show the water meets the required standards, the water can be placed in the storage tanks for use in the plant or be released to the environment. If the samples show the water does not meet the standards, it will be reprocessed.

Some materials, such as the evaporator bottoms (solids that remain after the water is evaporated off), will be mixed with some material to form a solid (such as concrete). This is also sometimes done with spent demineralizer resins. After mixing with a hardener, the material is processed as solid radioactive waste. Gaseous wastes are filtered, compressed to take up less space, and then allowed to decay for some time period. After the required time has passed, the gases will be sampled. If the required limits are met, the gases will be released to the atmosphere, or sometimes the gases will be reused in specific areas of the plant. Solid wastes are packaged as required and shipped to a burial site for disposal.

Final disposal of high-level waste is delayed for 40-50 years to allow its radioactivity to decay, after which less than one thousandth of its initial radioactivity remains, and it is much easier to handle. Hence canisters of vitrified waste, or used fuel assemblies, are stored under water in special ponds, or in dry concrete structures or casks, for at least this length of time.

The ultimate disposal of vitrified wastes, or of used fuel assemblies without reprocessing, requires their isolation from the environment for a long time. The most favored method is burial in stable geological formations some 500 meters deep. Several countries are investigating sites that would be technically and publicly acceptable, and in Sweden and Finland construction is proceeding in 1.9 billion year-old granites.

One purpose-built deep geological repository for long-lived nuclear waste (though only from defense applications) is already operating in New Mexico, in a salt formation. After being buried for about 1000 years most of the radioactivity will have decayed. The amount of radioactivity then remaining would be similar to that of the corresponding amount of naturally-occurring uranium ore from which it originated, though it would be more concentrated.

Layers of Protection after Disposal

To ensure that no significant environmental releases occur over a long period after disposal, a 'multiple barriers' disposal concept is used. The radioactive elements in high-level (and some intermediate-level) wastes are immobilized and securely isolated from the biosphere. The principal barriers are:

a) Immobilize waste in an insoluble matrix, e.g. borosilicate glass (or leave them as uranium oxide fuel pellets - a ceramic).

b) Seal inside a corrosion-resistant container, e.g. stainless steel.

c) Surround containers with bentonite clay to inhibit any groundwater movement if the repository is likely to be wet.

d) Locate deep underground in a stable rock structure.

e) For any of the radioactivity to reach human populations or the environment, all of these barriers would need to be breached, and this would need to happen before the radioactivity decayed to innocuous levels.

REFERENCES

Albritton, Daniel, "What Should Be Done in a Science Assessment In Protecting the Ozone Layer: Lessons, Models, and Prospects," 1998.

Allied Signal Corporation. "Remarks," International CFC and Halon Alternatives Conference. Washington, DC. 1989.

Alternative Fluorocarbons Environmental Acceptability Study (AFEAS), Washington, DC, 1995. "Production, Sales, and Atmospheric Release of Fluorocarbons,"

Alternative Fluorocarbons Environmental Acceptability Study (AFEAS), Washington, DC 1996.

Andelin and John, "Analysis of the Montreal Protocol," Staff report, U. S. Congress, Office of Technology Assessment, Jan. 13, 1988.

Angell, J. K., and J. Korshover, "Quasi-biennial and Long-term Fluctuations in Total Ozone," Monthly Weather Review vol. 101, pp.426–43, 2005.

Anderson, James G. "The Measurement of Trace Reactive Species in the Stratosphere: An Overview." In Causes and Effects of Stratospheric Ozone Depletion: An Update, Washington, DC: National Academy Press, 2008.

Angell, J. K. "The Variations in Global Total Ozone and North Temperate Layer Mean Ozone." Journal of Applied Meteorology, vol. 27, no. 1, pp. 91–97, 2007.

Clark , B.D, Gilard, A, Bisset, R. and Tomlinson, R. 1984 perspectives on environmental impact assessment. Reidel publishing company, Holland.

Chunmei, Wang; Zhaolan, Lin. (2010). "Environmental Policies in China over the past 10 Years: Progress, Problems and Prospects". International Society for Environmental Information Sciences 2010 Annual Conference (ISEIS) 2: 1701–1712.

Carle, Mark. A., Mickey Sarquis, and Louise Mary Nolan. 1991. Physical Science: The Challenge of Discovery. D.C. Health and Company, Lexington, Massachusetts.

Christensen, John W. 1991. Global Science Laboratory Manual. Kendall/Hunt Publishing Company, Dubuque, Iowa.

D. H. Stedman, "Atomic Chlorine and the Chlorine Monoxide Radical in the Stratosphere: Three in Situ Observations." Science, vol.198, 1981.

Dubey, S, Newnes, D, 2003. Green democracy peoples participation in environmental decision making, environmental justice initiative.

Grove, R.H. (1995) Green Imperialism: Colonial Expansion, Tropical Island Edens, and the Origins of Environmentalism, 1600-1860 New York: Cambridge University Press.

Govind, R. and Dolloff F. Bishop, "Biofiltration for Treatment of Volatile Organic Compounds (VOCs) in Air", Chapter in Biodegradation Technology Developments, Volume II, Eds. Subhas K. Sikdar and Robert L. Irvine, Technolmic Publishing Company, Lancaster, PA (1998).

Govind, R. and Zhao Wang, "Effect of Support Media on Is0-pentane Biofiltration", Paper submitted to Environmental Progress (1998).

Govind, R., Fang, J., R. Melarkode, "Biotrckling Filter Pilot Study for Ethanol Emissions Control", A Report prepared for the Food Manufacturing Coalition for Innovation and Technology Transfer, by PRD TECH, Inc., Florence, KY (1998).

Hocking, Colin, et al. 1990. Global Warming and the Greenhouse Effect. Lawrence Hall of Science, University of California at Berkeley, Berkeley, California.

Kulp, J.L. and Herrick, C.N. The Causes and Effects of Acid Deposition. Interim Assessment, National Acid Precipitation Assessment Pro- gram. Washington, D.C.: U.S. Government Printing Office, 1987.

Kosteltz, A.M., A. Finkelstein and G. Sears, "What are the real opportunities in biological gas cleaning for North America, in Proceedings of the 89th Annual Meeting and Exhibition of the Air and Waste Management Association, Air and Waste Management Association, Pittsburgh, PA (1996).

Michael Northcott, A Moral Climate: the ethics of global warming, 2007 Intergovernmental Panel on Climate Change, Climate Change 2007: The Physical Science Basis Summary for Policy Makers Millennium Ecosystem Assessment.

Miller, Tyler G. 1990. Living in the Environment. Wadsworth Publishing Co., Belmont, California.

Morrisette, Peter M. "The Evolution of Policy Responses to Stratospheric Ozone Depletion". Natural Resources Journal, vol. 29, 1995.

Mohnen, V.A. "The Challenge of Acid Rain:' Sci- entific American, Vol. 259, No. 2 (1988).

Roa, Michael L. 1993. Environmental Science Activities Kit. The Center for applied Research in Education, Professional Publishing, West Nyack, New York.

Stephen O., E. Thomas Morehouse, Jr., and Alan Miller, "The Military's Role in Protection of the Ozone Layer." Environmental Science and Technology, vol 28, no. 13, 1994.

Yudelson, J.M., "The future of the U.S. biofiltration industry, in Proceedings of the 1996 Conference on Biofiltration (an Air Pollution Control Technology), Reynolds, F.E., Ed., The Reynolds Group, Tustin, CA, page 1 (1996).

8

Biopesticides and Integrated Pest Management

Pesticides are substances meant for preventing, destroying or mitigating any pest. They are a class of biocide. The most common use of pesticides is as plant protection products (also known as crop protection products), which in general protect plants from damaging influences such as weeds, diseases or insects. This use of pesticides is so common that the term pesticide is often treated as synonymous with plant protection product, although it is in fact a broader term, as pesticides are also used for non-agricultural purposes. The term pesticide covers a wide range of compounds including insecticides, fungicides, herbicides, rodenticides, molluscicides, nematicides, plant growth regulators and others.

A pesticide is generally a chemical or biological agent (such as a virus, bacterium, antimicrobial or disinfectant) that through its effect deters, incapacitates, kills or otherwise discourages pests. Target pests can include insects, plant pathogens, weeds, mollusks, birds, mammals, fish, nematodes (roundworms), and microbes that destroy property, cause nuisance, spread disease or are vectors for disease. Although there are benefits to the use of pesticides, some also have drawbacks, such as potential toxicity to humans and other animals. According to the Stockholm Convention on Persistent Organic Pollutants, 9 of the 12 most dangerous and persistent organic chemicals are pesticides. Many pesticides can be grouped into chemical families. Prominent insecticide families include Organochlorine, organophosphates, and carbamates. Organochlorine hydrocarbons (e.g. DDT) could be separated into dichlorodiphenylethanes, cyclodiene compounds, and other related compounds.

EFFECTS OF USE OF PESTICIDES

Pesticides can travel great distances through the environment. When sprayed on crops or in gardens, pesticides can be blown by the wind to other areas. They can also flow with rain water into nearby streams or can seep through the soil into ground water. Some pesticides can remain in the environment for many years and pass from one organism to another. Pesticides differ according to their effects on various organisms. Selective pesticides are toxic only to the target pests. They cause little or no harm to other organisms **(Fig 8.1)**. However, nonselective pesticides can harm or even kill organisms that are not considered pests.

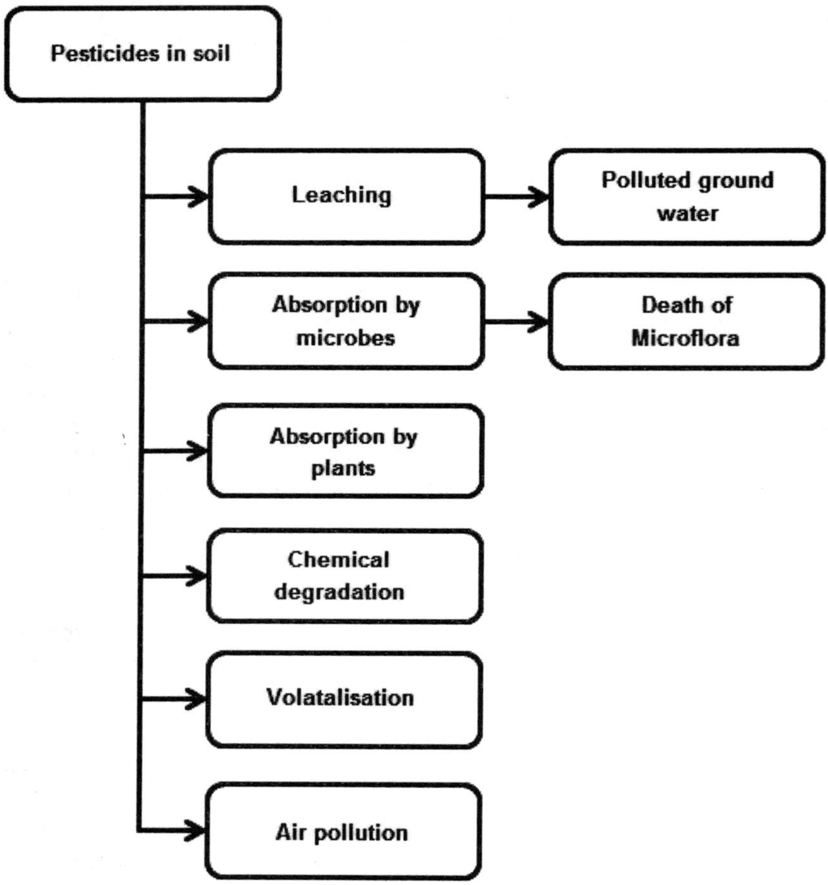

Fig. 8.1: Effect of use of pesticides

A variety of synthetic insecticides have been evaluated which generally affect non-target organisms (where there is no target to kill the non-pathogens). As a result of their use in agriculture, many beneficial organisms are killed. In turn, they cast hazardous effects on man also. Pesticide poisonings of people, livestock, and wildlife have occurred when proper care was not taken. Pesticide applicators must be very careful to avoid these risks. Mishandling of pesticides can lead to:

(i) Reduced control of the target.

(ii) Injury of non-target plants and animals.

(iii) Environmental damage.

(iv) May disrupt the natural balance in ecosystems.

The amount of pesticide that migrates from the intended application area is influenced by the particular chemical's properties: its propensity for binding to soil, its vapor pressure, its water solubility, and its resistance to being broken down over time. Factors in the soil, such as its texture, its ability to retain water, and the amount of organic matter contained in it, also affect the amount of pesticide that will leave the area. Some pesticides contribute to global warming and the depletion of the ozone layer.

Effect on Air

Pesticides can contribute to air pollution. Pesticide drift occurs when pesticides suspended in the air as particles are carried by wind to other areas, potentially contaminating them. Pesticides that are sprayed on to fields and used to fumigate soil can give off chemicals called volatile organic compounds, which can react with other chemicals and form a pollutant called tropospheric ozone. Pesticide use accounts for about 6 percent of total tropospheric ozone levels.

Effect on Water

There are four major routes through which pesticides reach the water: it may drift outside of the intended area when it is sprayed, it may percolate, or leach, through the soil, it may be carried to the water as runoff, or it may be spilled, for example accidentally or through neglect. Water soluble pesticides are easily transported from the target area into ground water and streams since the pesticides get dissolved in the water. This means that the pesticides may run off to other areas and cause damage to un-targeted animals and plants in other places.

Effect on Soil

The use of pesticides decreases the general biodiversity in the soil. Many of the chemicals used in pesticides are persistent soil contaminants, whose impact may endure for decades and adversely affect soil conservation. Depending on the chemical nature of the pesticide, degradation and sorption control directly the transportation from soil to water, and in turn to air and our food.

Effect on Plants

Nitrogen fixation, which is required for the growth of higher plants, is hindered by pesticides in soil. The insecticides DDT, methyl parathion, and especially pentachlorophenol have been shown to interfere with legume-rhizobium chemical signaling. Reduction of these symbiotic chemical signaling results in reduced nitrogen fixation and thus reduced crop yields. Pesticides can kill various insects which help in pollination. Pesticides have some direct harmful effect on plant including poor root hair development, shoot yellowing and reduced plant growth. They can also kill non target plants and enter into the plant tissue which in turn causes health issues with man and other animals.

Effect on Animals

Animals may be poisoned by pesticide residues that remain on food after spraying, for example when wild animals enter sprayed fields or nearby areas shortly after spraying. Poisoning from pesticides can travel up the food chain; for example, birds can be harmed when they eat insects and worms that have consumed pesticides. Fat soluble pesticides are readily absorbed in insects, fish, and other animals, often resulting in extended persistence in food chains.

Bioaccumulation refers to the accumulation of substances, such as pesticides, or other organic chemicals in an organism. Bioaccumulation occurs within a trophic level, and is the increase in concentration of a substance in certain tissues of organisms' bodies due to absorption from food and the environment. Bioconcentration is defined as occurring when uptake from the water is greater than excretion. DDT is thought to biomagnify and biomagnification is one of the most significant reasons it was deemed harmful to the environment by the EPA and other organizations. DDT is stored in the fat of animals and takes many years to break down, and as the fat is consumed by predators, the amounts of DDT biomagnify.

Effect on Humans

Health effects of pesticides may be acute or delayed in workers who are exposed. Strong evidence also exists for negative outcomes from pesticide exposure including neurological, birth defects, fetal death, and neuro-development disorder. There are two broad categories of health effects caused by pesticides, short-term (or acute) effects and long-term (or chronic) effects.

The short-term (or acute) effects vary depending on the chemical, the dose received and the susceptibility of the individual exposed. If people don't follow instructions on how to avoid over-exposure, they can experience burning, stinging or itching of eyes, nose, throat and skin. Other acute symptoms may include nausea, vomiting, diarrhea, wheezing, coughing and headache.

Medical science is less certain about the long-term health effects that may result from routine exposures, such as might occur when someone steps out into their yard after pesticides have been applied or when children play on treated turf, especially when these exposures are repeated over time.

(i) Researchers suggest that there is a moderately elevated risk of reproductive or developmental effects from direct exposures to some pesticides prior to conception, or during prenatal or postnatal periods. Although not all studies are consistent, several studies of women who work with pesticides suggest that some pesticides are associated with fertility problems and increased risks of spontaneous abortion and miscarriage.

(ii) Several pesticides are associated with effects on the central or peripheral nervous systems of animals and humans. There is mounting evidence suggesting a moderate, but significant, elevation in the risk of neurodegenerative disorders, such as Parkinson's disease (PD), among those persons occupationally exposed to pesticides, particularly farmers30 and gardeners.

(iii) Many studies have examined the effects of pesticide exposure on the risk of cancer. Associations have been found with: leukemia, lymphoma, brain, kidney, breast, prostate, pancreas, liver, lung, and skin cancers. Increased rates of cancer have been found among farm workers who apply these chemicals. A mother's occupational exposure to pesticides during pregnancy is associated with increase in her child's risk of leukemia and brain cancer.

The use of chemical pesticide and other agro chemicals are getting reduced /being banned globally because of their toxic effects on human beings and his live stock, residual toxicity, environmental problems, pest out-breaks and drastic effects on beneficial insects. Therefore, now it is imperative to develop a holistic system of tackling pests to make it more eco-friendly, economically viable and socially acceptable for the farmers. In this regard to tackle the major pests and diseases of major crops biotechnological approaches are gaining momentum.

BIOPESTICIDES

Biopesticides are certain types of pesticides derived from such natural materials as animals, plants, bacteria, and certain minerals. Biopesticides, a contraction of 'biological pesticides', include several types of pest management intervention: through predatory, parasitic, or chemical relationships. The term has been associated historically with biological control - and by implication - the manipulation of living organisms. They are typically created by growing and concentrating naturally occurring organisms and/or their metabolites including bacteria and other microbes, fungi, nematodes, proteins, etc. They are often considered to be important components of integrated pest management (IPM) programmes, and have received much practical attention as substitutes to synthetic chemical plant protection products (PPPs).

Benefits of Biopesticides

Biopesticides provide a wide range of benefits to growers, packer/shippers, food processors, and food retailers, as well as consumers **(Fig 8.2)**. In addition to food uses, biopesticides may also contribute protection of turf, ornamentals, and forests. They are also useful in the realm of public health where they can be used for disease and nuisance management, i.e. mosquito and tick control. While human health benefits tend to be indirect, the benefits of using biopesticides in agricultural application tend to be more direct. The various benefits of biopesticides can be summarized as follows:

Crop Quality and Yield

Biopesticides are often included as "standard" inputs in production programs as a means to significantly improve quality and yield of crops under challenging conditions. Such products, which include a variety of both microbial and biochemical biopesticides, may be used to protect the crop from pathogens, insect pests, and/or weeds that can divert or restrict the crop's access to valuable resources such as water, sunlight, or nutrients. Thus, biopesticides help to promote crop health thereby increasing its salability.

Plant growth regulators (PGRs) enable growers to improve crop quality and yield in a different way. Rather than mitigating pest damage to the crop, PGRs evoke physiological benefits such as increased fruit size or enhanced color. As color and fruit size are often key determinates in the price a grower receives for a crop, application of biopesticides have the ability to increase overall yield resulting in higher net farm income.

Effect on Ecosystem

Biopesticides target specific pests without disrupting the beneficial components of an agro ecosystem. For instance, most bioinsecticides do not harm beneficial insect populations and bioherbicides tend to be specific for a single weed species. Still, some biopesticides, particularly those used for disease management, tend to be broader spectrum, making them useful for managing the multiple pathogens that challenge the growth of most crops.

While many insect species have a negative impact on crop production, beneficial insects have the opposite effect. These benefits may derive from the insect's role as a pollinator or through its place as a natural predator to the insects that cause damage. Several species of beetles, flies, and wasps fall into this category. By helping to maintain populations of natural enemies to damaging insects, biopesticides play an important role in integrated pest management programs.

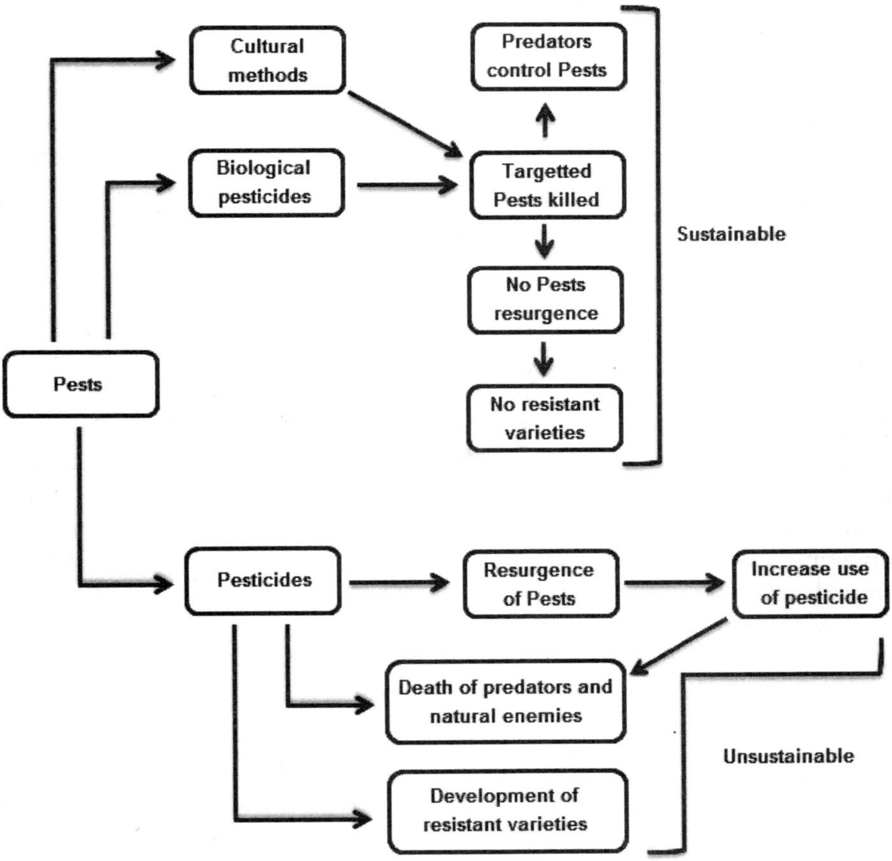

Fig. 8.2: Comparative account of chemical pesticides and biopesticides

Labor and Harvest Flexibility

Qualities shared by most biopesticides provide growers with more options to maximize their labor force and optimize harvest times. Biopesticides with short Pre-Harvest Intervals (PHIs) allow harvest and shipping schedules to be better maintained after required pesticide applications.

Plant Growth Regulators (PGRs), can provide harvest flexibility in a different way. Some PGRs inhibit the biosynthesis of ethylene in plants. As a natural plant hormone involved in fruit maturation, ripening and abscission, ethylene is a key contributor in a crop's "readiness" for harvest. So, ethylene inhibitors can help growers with harvest scheduling as well as the maintenance of fruit quality during storage.

Integrated Pest Management (IPM) Compatibility

The IPM approach combines cultural, biological and chemical means to control pests, all the while minimizing economic, public health and environmental risks. Biopesticides are considered among the best low-risk and most highly effective tools for achieving crop protection in IPM systems. Research, field trials, and performance history prove that the most effective IPM programs typically consist of biopesticides used in combination or rotation with traditional chemistries. This optimizes the grower's ability to:

(i) Successfully manage pests with a variety of effective control mechanisms.

(ii) Manage pesticide resistance through rotation of these effective chemistries.

(iii) Minimize the environmental impact of the production system.

Resistance Management

Populations of insect pests, plant pathogens and weeds all have the ability to develop resistance quickly, even to different types of functionally similar chemistries. This phenomenon is called cross-resistance and is caused by multi-chemistry detoxification mechanisms present in many pest populations. Biopesticides have long been used in combination with synthetic chemistries to provide the basis for excellent control programs that effectively manage resistance. Biopesticides typically have modes of action that are unique from synthetic pesticides and do not rely on a single target site for efficacy. The naturally occurring soil bacterium Bacillus thuringiensis (Bt), for example, has multiple active components including a range of different toxins and germinating spores. Properly used, biopesticides have the potential to extend the effective field life of all products by curtailing the development of resistant pest populations.

Environmental Safety

Biopesticides provide growers with valuable tools on both fronts by delivering solutions that are highly effective in managing pests, without creating negative impacts on the environment. Overall, biopesticides have very limited toxicity to birds, fish, bees and other wildlife. They help to maintain beneficial insect populations, break down quickly in the environment, and may serve to reduce conventional pesticide applications through their effective use in resistance management programs.

Applications of Biopesticides

Biopesticides are typically microbial biological pest control agents that are applied in a manner similar to chemical pesticides. In order to implement these environmentally friendly pest control agents effectively, it can be important to pay attention to the way they are formulated and applied. Various applications **(Fig 8.3)** are as follows:

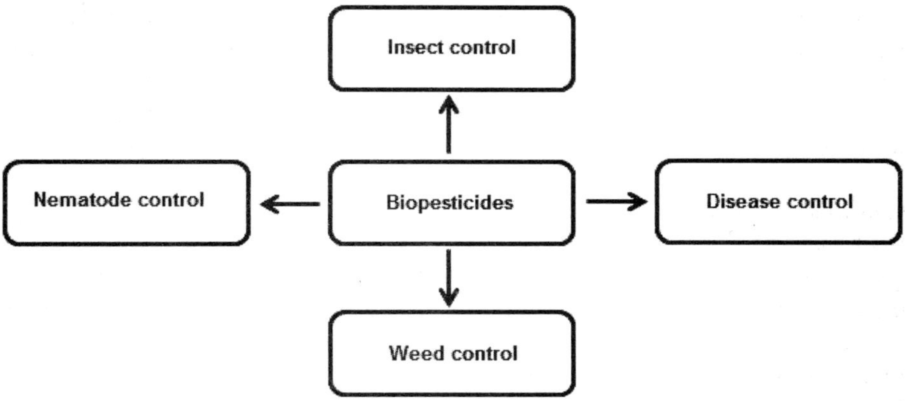

Fig. 8.3: Applications of biopesticides

Insect Control

Managing insect pests in ways that leave little or no toxic residues, have minimal impact on non-target organisms, and are not prone to pest counter-adaptation (resistance) has always been challenge in modern agricultural systems. Some microbial biopesticides are used to kill mosquito larvae without contaminating the water in which they live. Such products have been proven to be a valuable and environmentally-friendly tool in public health programs to limit the spread of malaria, yellow fever, and other human diseases transmitted by mosquitoes. And insect-parasitic nematodes (microscopic roundworms) are both highly specific and effective as a means for controlling soil-dwelling weevil larvae infesting citrus tee root systems. Use of bioinsecticides also helps to extend the useful life of synthetic insecticides and to reduce the amount of unwanted pesticide residues in vegetable and fruit crops.

Disease Control

Biopesticides, key components of integrated pest management (IPM) programs, are receiving much practical attention as a means to reduce the load of synthetic chemical products being used to control plant diseases. Biopesticides for use against crop diseases have already established themselves on a variety of crops. For example, biopesticides already play an important role in controlling downy mildew diseases.

Weed Control

Weeds present a multifaceted problem in agriculture. Weeds are responsible for reducing crop yields by competing for space, sunlight, nutrients, and water. Weeds may also function as alternate hosts for pest insects and plant diseases that impact crop growth, yields, and quality. The use of biopesticides for weed control presents a difficult challenge due to the physiological similarities between crop plants and weed species. In recent years, however, scientists have identified several disease causing organisms that specifically and effectively attack key weed species. And, some plant extracts have also been identified that have broad spectrum herbicidal activity against numerous weed species. Biopesticides based on such active ingredients can be used to reduce our dependency on chemical herbicides.

Nematode Control

Plant parasitic nematodes have a stylet, or mouth-spear, similar to a hypodermic needle. The stylet is used to puncture plant cells and inject digestive enzymes and other fluids. The nematode then draws plant fluids through the stylets. Nematologists have identified several bacterial and fungal products for control of soil nematodes, as well as plant extracts that display nematoxicity or control the pest indirectly by boosting the natural defenses of crop plants. As fumigant nematicides continue to be phased out, biopesticides will play an ever-increasing role in control of these important pests.

Types of Biopesticides

Biopesticides fall into three major classes (Fig 8.4):

(i) Microbial biopesticides: Microbial pesticides consist of a microorganism (e.g., a bacterium, fungus, virus or protozoan) as the active ingredient. Microbial pesticides can control many different kinds of pests, although each separate active ingredient is relatively specific for its target pests. For example, there are fungi that control certain weeds, and other fungi that kill specific insects.

(ii) Plant-Incorporated-Protectants (PIPs): Plant-Incorporated-Protectants (PIPs) are pesticidal substances that plants produce from genetic material that has been added to the plant. For example, scientists can take the gene for the Bt pesticidal protein, and introduce the gene into the plant's own genetic material. Then the plant, instead of the Bt bacterium, manufactures the substance that destroys the pest.

(iii) Biochemical pesticides: Biochemical pesticides are naturally occurring substances that control pests by non-toxic mechanisms. Biochemical pesticides include substances, such as insect sex pheromones that interfere with mating as well as various scented plant extracts that attract insect pests to traps.

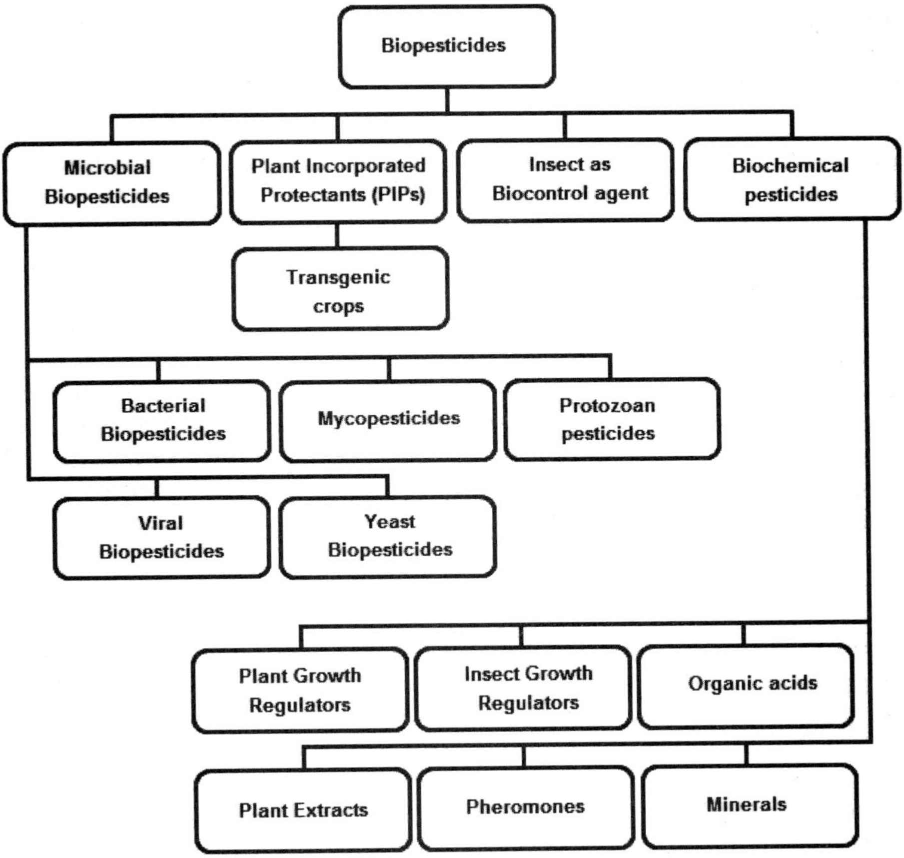

Fig. 8.4: Types of Biopesticides

Microbial Biopesticides

Microbial biopesticides are products derived from various microscopic organisms. Microbial products may consist of the organisms themselves and/or the metabolites they produce. There is a vast majority of microorganisms such as viruses, bacteria, fungi, protozoa, and mycoplasma, known to kill the insect pests. The suitable preparations of such microorganisms for control of insects are called as "microbial insecticides". The microbial insecticides are non-hazardous, non-phytotoxic and selective in their action. Pathogenic microorganisms which kill insects are viruses (DNA containing viruses e.g. baculovirus iridovirus, entomopoxvirus), bacteria (*Bacillus thuringiensis, B. popilliae, B. sphaericus, B. moritai*) and fungi (e.g. *Aspergillus, Coelomomyces, Entomophthora, Fusarium, Hirsutella, Paecilomyces*).

Bacterial Pesticides

Biopesticides based on bacteria have been used to control plant diseases, nematodes, insects, and weeds. Bacteria are present in all soils and are the most abundant micro-organisms in soil samples. Many spore forming and non-spore forming bacteria are known to be effective against a wide spectrum of insects and diseases.

The most well-known and widely used of all biopesticides are insecticides based on *Bacillus thuringiensis*, commonly referred to as "Bt." During spore formation, Bt produces insecticidal proteins (know as delta-endotoxins) that kill caterpillar pests , fly and mosquito larvae , or beetles (depending on the subspecies and strain of Bt) that ingest them through feeding in Bt-treated areas. The highly specific delta-endotoxins bind to and destroy the cellular lining of the insect digestive tract, causing the insect to stop feeding and die. *Bacillus sphaericus* is another insecticidal bacterium that has been used successfully to control certain mosquito species. Other bacteria are used for the control of plant pathogens. Certain strains of *Bacillus subtilis*, *Bacillus pumilus*, *Pseudomonas spp.*, and *Streptomyces spp.* increase yield and prevent plant diseases by outcompeting plant pathogens in the rhizosphere, producing anti-fungal compounds, and by promoting plant and root growth.

Mycopesticides

Different fungal biopesticides can be used to control plant diseases (caused by other fungi, bacteria or nematodes), as well as some insect pests and weeds. Because fungal biopesticides are so diverse in nature, their means of affecting the target pest can be equally diverse. The most common modes of action are through competitive exclusion, mycoparasitism, and production of metabolites. Some fungi can exhibit all of these modes of action. The infective propagules e.g. conidia, spores, etc. of the antagonistic fungi reach the haemocoel of the insect either through integument or mouth. They get attached to epicuticle, germinate and penetrate the cuticle either by germ tubes or infection peg. They multiply in haemocoel followed by secretion of mycotoxins which result in death of insect hosts.

In England, *Verticillium lecanii* has shown as a potential antagonist against aphid pests affecting Chrysanthemum in green houses. In USSR, spraying of aphids and spider mites with *Entomophthora thaxteriana* and *E. sphaerosperma* resulted in 95 per cent mortality within 24 h by secreting the mycotoxins. *E. thaxteriana* suspension when applied on aphids of apple trees resulted in about 74 per cent mortality without harmful effects on natural predators.

Trichoderma spp. is some of the most common fungi in nature. Many beneficial Trichoderma have the ability to readily colonize plant roots, without harming the plant. It is this close relationship with the plant that makes these species excellent biocontrol agents. These microbial biofungicides can out-compete pathogenic fungi for food and space, and in the process can stimulate plant host defenses and affect root growth. In addition, they have the ability to attack and parasitize plant pathogens under certain environmental conditions.

Beauveria bassiana is a fungus that acts as a parasite on many insect species. *B. bassiana* has a broad host range, although individual strains may be restricted in the number of insects it can attack. Its spores adhere directly to the host cuticle, where they will germinate, produce enzymes that attack and dissolve the cuticle, penetrate, and grow into the insect's body, feeding on internal tissues and releasing an insect toxin. As the insect dies, it changes color to pink or brown and eventually the entire body cavity is filled with fungal mass. *B. bassiana* has proved effective in controlling troublesome crop pests such as aphids, thrips and whitefly – even chemical pesticide-resistant strains such as Q-Biotype Whitefly.

Protozoa as Biopesticides

Protozoa are single-celled eukaryotic organisms that exist in both water and soil. While most protozoa feed on bacteria and decaying organic matter, a wide range of protozoan species are insect parasites.

The protozan *Nosema locustae* is known to be a natural biocontrol agent of many grasshopper species. Nosema infects at least 90 species of grasshoppers. It is non-toxic to humans and other mammals, as well as the over 250 natural predators of grasshoppers. After consuming Nosema locustae, grasshopper feeding is typically curtailed within a week. Within two weeks or longer, as many as 50% of the infected insects die and approximately half of the surviving population remains weak, consuming 75% less forage than a healthy insect. An important function in the transmission of Nosema spores to healthy grasshoppers occurs as the insects scavenge and cannibalize infected cadavers. Since infected grasshoppers develop a large number of the spores within them, the cannibalizing grasshoppers get a much greater dose of the disease causing organism than through the initial Nosema application.

Viral Biopesticides

Preparations of viruses or their products have been developed as effective biopesticide and being successfully used for the control of insect pests in agriculture, forestry and horticulture. This method of disease control is free from pollution, toxicity or any hazards related to plant or animal health. However, these viruses are specific and have no harmful effects on useful insect pollinators, insects yielding useful products, warm blooded animals and even man. After application viruses get entered into the mouth and digestive tract of insect pest and kill them.

Baculoviruses are a family of naturally-occurring viruses known to infect only insects and some related arthropods. Most are so specific in their action that they infect and kill only one or a few species of Lepidoptera larvae (caterpillars), making them good candidates for management of crop pests with minimal off-target effects. Baculoviruses used as microbial biopesticides consist of DNA surrounded by a protein coat (nucleocapsid), which is itself embedded in a protein "microcapsule" or occlusion body (OB) that provides some protection from degradation in the environment. Depending on the virus, OBs may contain a single nucleocapsid (a granulovirus, or GV) or multiple nucleocapsids (nucleopolyhedrovirus, or NPV).

Nuclear polyhedrosis viruses (NPVs) which belong to subgroup of Baculoviruses have been used for the preparation of potential pesticides. Heliothis sp. is a cosmopolitan insect pest attacking at least 30 food and fiber yielding crop plants. They have been controlled by application of NPVs of Baculovirus heliothis. Lymantria dispar, commonly known as gypsy moth, is a serious pest of forest trees. It has been successfully controlled by gypsy moth Baculovirus i.e. NPV preparations.

The granulovirus of the codling moth *Cydia pomonella,* or CpGV, is a good example of a commercially successful viral insecticide. First discovered in the 1960 s, it is now the active ingredient of about a half-dozen products sold worldwide. Often used in conjunction pheromone-based mating disruption, CpGV limits codling moth populations and damage in pome fruits while preserving beneficial insects and minimizing chemical residues. Although accepted for use in organic farming, most CpGV applications occur in conventional orchards where its unique mode of action can minimize risk of resistance to chemical insecticides.

There is another group of virus, CPVs, which are present in more than 200 insects, out of which only a few have been used for control of insect pests. For the first time in France, a CPV of pine processionary caterpillar (Thaumetopoea pityocampa) was applied as pesticide used for the control of pine forest pests.

Yeast as Biopesticides

A variety of yeasts have been investigated for their usefulness in controlling plant diseases. Non-pathogenic Cryptococcus and Candida species naturally occur on plant tissues and in water. Isolates from a variety of crops have been investigated for their biocontrol capacities. For example, Candida oleophila Strain O, first isolated from golden delicious apples, has been developed into an effective biopesticide for the control of post-harvest fruit rots. It is applied to apples and pears after harvest — but before storage — to control particular fungal pathogens. The yeast serves as an antagonist to fungal pathogens such as gray mold (Botrytis cinerea) and blue mold (Penicillium expansum), which cause post-harvest decay.

Candida oleophila Strain O works primarily through competition for nutrients and pre-colonization of plant wound sites. However, there is evidence that it produces enzymes that can degrade fungal cell walls and stimulate plant host defense pathways in freshly harvested fruit, both of which may also make a restricted contribution to the strain's antagonistic activity.

Insects as Biocontrol Agents

Insects play a valuable role in the eradication of valuable weeds but complete eradication has not been achieved by this method. Insects attack the host and multiply rapidly until much plant growth is destroyed. Due to food shortage the number of insects decreases, consequently weeds reappear. Chinese used predator ants to control certain insect pests of citrus. This practice continued through the ages, and even in modern time, it is still in use. Some other examples are:

(i) *Senecio jacobaea*, a serious weed in wetter areas has been controlled by introducing Tyria jacobaea, cinnaber moth, in California (U.S.A.) and by Pegohylemyia seneciella in Australia.

(ii) *Hypericum perforatum* (goat weed), a serious weed found in California and pacific North-West is controlled by Chrysolina hyperici (goat weed beetle).

(iii) *Eichhornia crassipes* (water hyacinth), is controlled by an insect, Neochetina eichhorniae. This insect was introduced from Latin America to India. The female lays eggs on petiole of water hyacinth. Larvae feed upon petiole and the adults feed on leaves resulting in destruction of the whole plant.

(iv) Lantana, a pasture pest and poisonous plant is found throughout the world. In Hawaii, caterpillars of Plusia verticillata were introduced for its control. Larvae of seedfly, *Agromyza lantanae*, eat many berries and cause others to dry so that birds could not carry them. The lace bug, Teleonemia scrupulosa was most effective insect to control Lantana but this insect could not yield much success.

Biochemical Biopesticides

Biochemical biopesticides are naturally occurring compounds or synthetically derived compounds that are structurally similar (and functionally identical) to their naturally occurring counterparts. In general, biochemical biopesticides are characterized by a non-toxic mode of action that may affect the growth and development of a pest, its ability to reproduce, or pest ecology. They also may have an impact on the growth and development of treated plants including their post harvest physiology. Biochemical biopesticides are divided into several different subcategories of products, including:

(i) Plant Growth Regulators

(ii) Insect growth Regulator

(iii) Organic Acids

(iv) Plant Extracts

(v) Pheromones

(vi) Minerals

Plant Growth Regulators

Plant Growth Regulators (PGRs) include a category of both natural and manufactured versions of natural substances that affect major physiological functions of plants. PGRs can promote, inhibit, or modify the physiological traits of a range of fruit, vegetable, ornamental and agronomic crops. There are five major classes of natural plant hormones. Under each class there are a number of PGR products that play specific and very important roles in optimizing crop yield and quality. They are:

a) Gibberellins

b) Cytokinins

c) Abscisic acid

d) Ethylene

e) Auxins

In addition to the major five classes of plant hormones, there are other plant compounds with suggested PGR activity that are involved with cell division, maturation, cell enlargement, plant defenses and plant resistance. These compounds include a variety of secondary metabolites produced by plants and some plant-associated microbes. Examples include polyamines, salicylic acid and signal peptides.

Insect Growth Regulators

Biopesticide Insect Growth Regulators (IGRs) have a unique mode of action separate from most chemical insecticides. Generally speaking, these products prevent insects from reaching a reproductive stage, thereby reducing the expansion of pest populations. IGRs can be divided into two broad categories; i.e. those that disrupt the hormonal regulation of insect metamorphosis, and those that disrupt the synthesis of chitin, a principal component of insect exoskeletons.

Azadirachtin is one of the most widely used botanical insect growth regulators. Because of its structural resemblance to the natural insect molting hormone ecdysone, Azadirachtin interrupts molting, metamorphosis, and development of the female reproductive system. Immature insects exposed to azadirachtin (mainly by ingestion) may molt prematurely or die before they can complete a properly timed molt. Those insects that survive a treatment are likely to develop into a deformed adult incapable of feeding, dispersing, or reproducing.

Organic Acids

Peracids are highly effective sanitizing agents used for control of pathogens and algae. Hydrogen peroxide is a well-known chemical compound that can be found in most people's medicine cabinets for disinfecting cuts and bruises, among other uses. Peracids combine hydrogen peroxide and organic acids, typically acetic acid, to form a new compound called peroxyacetic acid, abbreviated as PAA. This new compound is an activated form of hydrogen peroxide and produces a much more stable and powerful oxidizing compound to treat pathogens and algae in water.

The mode of action of peracids is by oxidation of cell membranes and penetration into cell structures of algae, bacteria and fungi. More specifically, peracids form free hydroxyl radicals (OH), which oxidize and disrupt thiol groups in proteins and enzymes.

Plant Extracts

Plants are, in effect, natural laboratories in which a great number of chemicals are biosynthesized. Many plants have developed natural, biochemical mechanisms to defend themselves from weed competition and animal, insect and fungal attacks. Some of these chemicals discourage feeding by insects and other herbivores. Others provide protection or even immunity from diseases caused by some pathogens. Still others help plants compete for resources by discouraging competition among different plant species. By studying the diverse chemistries of many different plant species, scientists have discovered many useful compounds that can be used as biopesticides.

Pheromones

Pheromones are chemical signals that trigger a natural response in another member of the same species. Insects release pheromones to serve many functions. These include secretion of pheromones to indicate the location of food sources, to warn others around about potential dangers, or locate a potential mate for reproduction.

Synthetic pheromones can be used in a number of ways to disrupt pest ecology and reduce crop damage. The first approach is to place small amounts of the female pheromone in lures to attract males into traps. Another, more common, use of synthetic pheromones is mating disruption. Growers saturate the environment with a sex pheromone so the male moths cannot easily locate females, disrupting their ability to mate. No insect mating means no fertilized eggs, and no larvae to damage crops.

Minerals

Minerals play a key role in a wide range of biopesticide applications that can be divided into three categories:

(i) Those that create barriers that keep pests from plant tissues and/or impact pathogen water supply.

(ii) Those that deliver physical impacts such as smothering or abrasion.

(iii) Those that act as an inert carrier for companion biopesticides.

Kaolin clay is a good example of a biopesticide that creates a physical barrier between insects and plant tissues. Kaolin also breaks off in small particles that attach to insects, agitating and repelling them. Potassium silicate is another example of this type of biopesticide, which also serves as a desiccant to soft bodies of insects and mites. Diatomaceous earth (DE) is an example of a mineral biopesticide that combats insect infestations through abrasion.

INTEGRATED PEST MANAGEMENT (IPM)

Integrated pest management (IPM), also known as Integrated Pest Control (IPC) is a broad based approach that integrates a range of practices for economic control of pests. Globalization of markets and increased movements of people all over the world are allowing for increasing numbers of invasive species to be brought into countries. Appropriate responses to these pests are needed and development and implementation strategies should be arranged. It is essential that the option that poses the least risks while maximizing benefits is needed and that the strategy may include all components related to integrated pest management strategies.

Integrated Pest Management (IPM) is an effective and environmentally sensitive approach to pest management that relies on a combination of common-sense practices. IPM programs use current, comprehensive information on the life cycles of pests and their interaction with the environment. This information, in combination with available pest control methods, is used to manage pest damage by the most economical means, and with the least possible hazard to people, property, and the environment. The IPM approach can be applied to both agricultural and non-agricultural settings, such as the home, garden, and workplace. IPM takes advantage of all appropriate pest management options including, but not limited to, the judicious use of pesticides. In contrast, organic food production applies many of the same concepts as IPM but limits the use of pesticides to those that are produced from natural sources, as opposed to synthetic chemicals.

Principles of IPM

IPM is not a single pest control method but, rather, a series of pest management evaluations, decisions and controls **(Fig 8.5)**. The various principles are:

Acceptable Pest Levels

IPM holds that wiping out an entire pest population is often impossible, and the attempt can be expensive and environmentally unsafe. IPM programmes first work to establish acceptable pest levels, called action thresholds, and apply controls if those thresholds are crossed. By allowing a pest population to survive at a reasonable threshold, selection pressure is reduced. This stops the pest gaining resistance to chemicals produced by the plant or applied to the crops. If many of the pests are killed then any that have resistance to the chemical will form the genetic basis of the future, more resistant, population. By not killing all the pests there are some un-resistant pests left that will dilute any resistant genes that appear.

Preventive Cultural Practices

Selecting varieties best for local growing conditions, and maintaining healthy crops, is the first line of defense, together with plant quarantine and 'cultural techniques' such as crop sanitation (e.g. removal of diseased plants to prevent spread of infection).

Monitor and Identify Pests

Not all insects, weeds, and other living organisms require control. Many organisms are innocuous, and some are even beneficial. IPM programs work to monitor for pests and identify them accurately, so that appropriate control decisions can be made in conjunction with action thresholds. This monitoring and identification removes the possibility that pesticides will be used when they are not really needed or that the wrong kind of pesticide will be used.

Mechanical Controls

Mechanical methods are the first options to consider. They include simple hand-picking, erecting insect barriers, using traps, vacuuming, and tillage to disrupt breeding.

Biological Controls

Natural biological processes and materials can provide control, with minimal environmental impact, and often at low cost. The main focus here is on promoting beneficial insects that eat target pests. Biological insecticides, derived from naturally occurring microorganisms (e.g.: Bt, entomopathogenic fungi and entomopathogenic nematodes), also fit in this category.

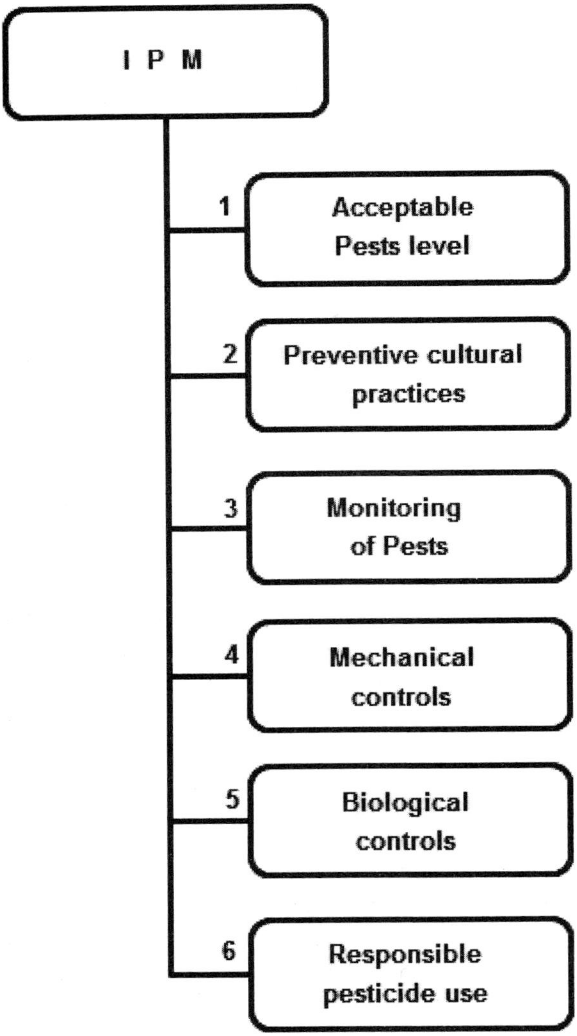

Fig. 8.5: Principles of Integrated Pest Management

Responsible Pesticide Use

Synthetic pesticides are generally only used as required and often only at specific times in a pest's life cycle. Many of the newer pesticide groups are derived from plants or naturally occurring substances (e.g.: nicotine, pyrethrum and insect juvenile hormone analogues), but the toxophore or active component may be altered to provide increased biological activity or stability.

IPM is the intelligent selection and use of pest control actions that will ensure favorable economic, ecological and sociological consequences and is applicable to most agricultural, public health and amenity pest management situations. Reliance on knowledge, experience, observation, and integration of multiple techniques makes IPM a perfect fit for organic farming. For large-scale, chemical-based farms (conventional), IPM can reduce human and environmental exposure to hazardous chemicals, and potentially lower overall costs of pesticide application material and labor.

REFERENCES

Alabouvette, C., Olivain, C. and Steinberg, C. (2006) Biological control of plant diseases: the European situation. European Journal of Plant Pathology 114, 329–341.

Altieri, Miguel A. 1994. Biodiversity and Pest Management in Agroecosystems. The Haworth Press, Binghamton, NY. 185 p.

Andersen, M.C., Ewald, M. and Northcott, J. (2005) Risk analysis and management decisions for weed biological control agents: ecological theory and modeling results. Biological Control 35, 330–337.

Anon. (2006) The Greening Waipara Project. Bio-Protection issue 2, November.

Ash, G.J. (2010) The science, art and business of successful bioherbicides. Biological Control 52, 230–240.

Asser-Kaiser, S., Fritsch, E., Undorf-Spahn, K., Kienzle, J., Eberle, K., Gund, N.A., Reineke, A., Zebitz, C.P.W., Heckel, D.G., Huber, J. and Jehle, J.A. (2007) Rapid emergence of baculovirus resistance in codling moth due to dominant, sexlinked inheritance. Science 318, 1916–1917.

Atawodi, S.E. and Atawodi, J.C. (2009) Azadirachta indica (neem): a plant of multiple biological and pharmacological activities. Phytochemistry Reviews 8, 601–620.

Bailey, K.L., Boyetchko, S.M. and Langle, T. (2010) Social and economic drivers shaping the future of biological control: a Canadian perspective on the factors affecting the development and use of microbial biopesticides. Biological Control 52, 221–229.

Bale, J.S., van Lenteren, J.C. and Bigler, F. (2008) Biological control and sustainable food production. Philosophical Transactions of the Royal Society B 363, 761–776.

Barratt, B.I.P., Howarth, F.G., Withers, T.M., Kean, J.M. and Ridley, G.S. (2010) Progress in risk assessment for classical biological control. Biological Control 52, 245–254.

Baxter, S.W., Chen, M., Dawson, A., Zhao, J.-Z., Vogel, H., Shelton, A.M., Heckel, D.G. and Jiggins, C.D. (2010) Mis-spliced transcripts of nicotinic acetylcholine receptor α6 are associated with fi eld evolved Spinosad resistance in Plutella xylostella (L.).

Bellows, T.S. and Fisher, T.W. (eds) (1999) Handbook of Biological Control. Academic Press, San Diego, California.

Bending, G.D., Aspray, T.A. and Whipps, J.M. (2008) Signifi cance of microbial interactions in the mycorrhizosphere. Advances in Applied Microbiology 60, 97–132.

Berg, G. (2009) Plant–microbe interactions promoting plant growth and health: perspectives for controlled use of microorganisms in agriculture. Applied Microbiology and Biotechnology 84, 11–18.

Benbrook, Charles M. 1996. Pest Management at the Crossroads. Consumers Union, Yonkers, NY. 272 p.

Flint, M.L. and van den Bosch, R. 1977. A Source Book on Integrated Pest Management. p. 173-174. Limited distribution. Supported by grant #G007500907 to UC International Center for Integrated and Biological Control.

Prakash, Anand and Jagadiswari Rao. 1997. Botanical Pesticides in Agriculture. CRC Press, Boca Raton, FL. 461 p.

Marschner, H. 1998. Soil-Root Interface: Biological and Biochemical Processes. p. 191-232. In: Soil Chemistry and Ecosystem Health. P.M. Huang (ed.). Soil Science Society of America, Inc., Madison, WI.

Norton, G.W. and J. Mullen. 1994. Economic Evaluation of Integrated Pest Management Programs: A Literature Review. Virginia Cooperative Extension Publication 448-120. 112 p.

http://www.biopesticideindustryalliance.org/benefitshome.php

http://www.jbiopest.com/users/LW8/page.php?intPageId=180

http://www.epa.gov/pesticides/biopesticides/whatarebiopesticides.htm

9

Biofuels and Biosensors

Energy powers the modern world. It is essential to the way we live, work and move today. Energy is essential in our daily lives. We use it to fuel our cars, grow our food, heat our homes, and run our businesses. Most of our energy comes from burning fossil fuels like petroleum, coal, and natural gas. These fuels provide the energy that we need today, but there are several reasons why we must develop sustainable alternatives. The supply of sustainable energy is one of the main challenges that mankind will face over the coming decades, particularly because of the need to address climate change. Energy can neither be created nor destroyed; it can be converted from one form to other. The non-renewable sources of energy like coal, petroleum, and natural gas will soon be finished, as there is continuous increasing demand for energy.

We are running out of fossil fuels. Fossil fuels take millions of years to form within the Earth. Fossil fuels are fuels formed by natural processes such as anaerobic decomposition of buried dead organisms. Scientists estimate the world will run out of fossil fuels within the next 50 to 120 years.

Moreover, use of fossil fuels can be harmful to humans and the environment. When fossil fuels are burned, they release carbon dioxide and other gases into the atmosphere. Some of these gases pollute the air we breathe and contribute to climate change – which threatens ecosystems and could lead to flooding, drought, or famine in some parts of the world. Therefore, there is need for clean and sustainable energy sources, which can meet the increasing demands of world. Under these circumstances renewable energy has good scope, due to following advantages over fossil fuels:

(i) Renewable energy technologies are clean sources of energy that have a much lower environmental impact than conventional energy technologies.

(ii) Renewable energy will not run out ever. Other sources of energy are finite and will someday be depleted.

BIOENERGY FROM BIOMASS

Biomass consists of any organic matter of vegetable or animal origin. It is available in many forms and from many different sources e.g. forestry products (biomass from logging and silvicultural treatments, process residues such as sawdust and black liquor, etc.); agricultural products (crops, harvest residues, food processing waste, animal dung, etc.); and municipal and other waste (waste wood, sewage sludge, organic components of municipal solid waste, etc). Biomass energy is solar energy stored in the chemical bonds of carbon and hydrogen chains as a result of photosynthesis or the metabolic activity of organisms. Biomass can be referred to as nature's solar battery reflecting its ability to store energy until required, which makes it more predictable and responsive than the sun or wind.

We have used biomass energy or bioenergy - the energy from organic matter - for thousands of years, ever since people started burning wood to cook food or to keep warm. And today, wood is still our largest biomass energy resource. But many other sources of biomass can now be used, including plants, residues from agriculture or forestry, and the organic component of municipal and industrial wastes. Even the fumes from landfills can be used as a biomass energy source. The use of biomass energy has the potential to greatly reduce our greenhouse gas emissions. Bioenergy is renewable energy made available from materials derived from biological sources. This widely available resource is receiving increased consideration as a renewable substitute for fossil fuels. If developed in sustainable manner and used efficiently, it can induce growth in developing countries, reduce oil demand, and address environmental problems. The potential benefits of Bio-Energy include: reduction of greenhouse gases, recuperation of soil productivity and degraded land, economic benefits from adding value to agricultural activities and improving access to and quality of energy services.

All managed sources of bioenergy will result in a net reduction in carbon dioxide emissions, if they replace coal-fired generation. For example, in the case of plantation timbers or crops (like sugar cane), the cycle of growing, harvesting and energy production does not produce or absorb any additional carbon. Carbon stored in the crop is released at harvest, and then reabsorbed by the next crop, similar to the natural carbon cycle **(Fig 9.1)**.

India is among top 10 countries in the world in terms of energy demand. As the country is on the fast track to growth, this demand is expected to grow exponentially. India imports 70% of its requirement of oil and is constantly threatened by the increasing prices of crude oil, uncertainty and environmental hazards that are connected with the consumption of fossil fuels. On the other hand, it is a predominantly agrarian economy, with availability of large quantities of biomass, which can be productively harnessed for meeting its burgeoning energy needs, as well as address larger developmental issues such as income and employment generation, environment protection and wasteland/saline land development, and improved quality of life of rural communities.

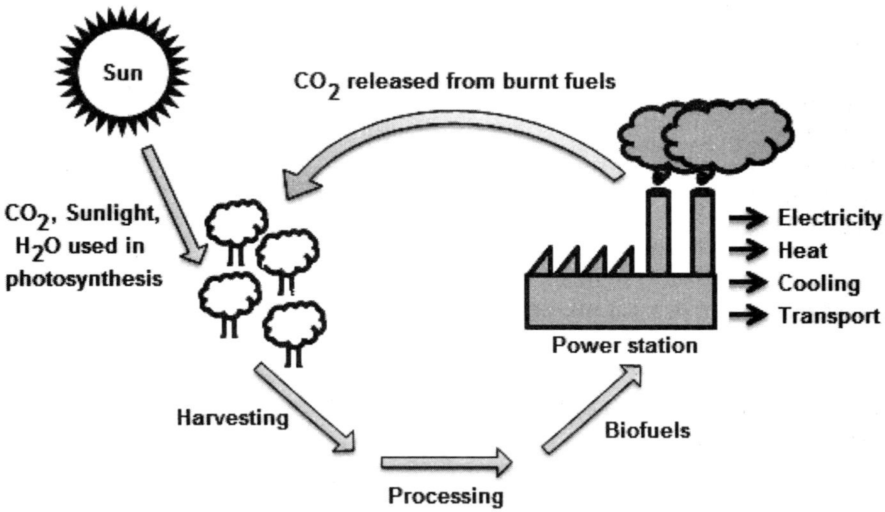

Fig. 9.1: Bioenergy production from biomass

There are three major biomass energy technology applications:

Biofuels

A biofuel is a fuel that uses energy from a carbon fixation. These fuels are produced from living organisms. Examples of this carbon fixation are plants and microalgae. These fuels are made from a biomass conversion. Biomass refers to recently living organisms, most often referring to plants or plant-derived materials. Biofuels have increased in popularity because of the rising oil prices and need for energy security. Unlike other renewable energy sources, biomass can be converted directly into liquid fuels - biofuels - for our transportation needs (cars, trucks, buses, airplanes, and trains). The two most common types of biofuels are ethanol and biodiesel.

Ethanol is an alcohol, the same found in beer and wine. It is made by fermenting any biomass high in carbohydrates (starches, sugars, or celluloses) through a process similar to brewing beer. Ethanol is mostly used as a fuel additive to cut down a vehicle's carbon monoxide and other smog-causing emissions. But flexible-fuel vehicles, which run on mixtures of gasoline and up to 85% ethanol, are now available.

Biodiesel is made by combining alcohol (usually methanol) with vegetable oil, animal fat, or recycled cooking greases. It can be used as an additive to reduce vehicle emissions (typically 20%) or in its pure form as a renewable alternative fuel for diesel engines.

Other biofuels include methanol and reformulated gasoline components. Methanol, commonly called wood alcohol, is currently produced from natural gas, but could also be produced from biomass. There are a number of ways to convert biomass to methanol, but the most likely approach is gasification. Gasification involves vaporizing the biomass at high temperatures, then removing impurities from the hot gas and passing it through a catalyst, which converts it into methanol. Most reformulated gasoline components produced from biomass are pollution-reducing fuel additives, such as methyl tertiary butyl ether (MTBE) and ethyl tertiary butyl ether (ETBE).

Biopower

Biopower, or biomass power, is the use of biomass to generate electricity. There are five major types of biopower systems: direct-fired, cofiring, gasification, anaerobic digestion, and pyrolysis.

Most of the biopower plants in the world use direct-fired systems. They burn bioenergy feed stocks directly to produce steam. This steam is usually captured by a turbine, and a generator then converts it into electricity. In some industries, the steam from the power plant is also used for manufacturing processes or to heat buildings. These are known as combined heat and power facilities. For instance, wood waste is often used to produce both electricity and steam at paper mills.

Many coal-fired power plants can use cofiring systems to significantly reduce emissions, especially sulfur dioxide emissions. Cofiring involves using bioenergy feed stocks as a supplementary energy source in high efficiency boilers.

Gasification systems use high temperatures and an oxygen-starved environment to convert biomass into a gas (a mixture of hydrogen, carbon monoxide, and methane). The gas fuels what's called a gas turbine, which is very much like a jet engine, only it turns an electric generator instead of propelling a jet.

The decay of biomass produces a gas - methane - that can be used as an energy source. In landfills, wells can be drilled to release the methane from the decaying organic matter. Then pipes from each well carry the gas to a central point where it is filtered and cleaned before burning. Methane also can be produced from biomass through a process called anaerobic digestion. Anaerobic digestion involves using bacteria to decompose organic matter in the absence of oxygen. Methane can be used as an energy source in many ways. Most facilities burn it in a boiler to produce steam for electricity generation or for industrial processes. Two new ways include the use of microturbines and fuel cells. Microturbines have outputs of 25 to 500 kilowatts. About the size of a refrigerator, they can be used where there are space limitations for power production. Methane can also be used as the "fuel" in a fuel cell. Fuel cells work much like batteries but never need recharging, producing electricity as long as there's fuel.

In addition to gas, liquid fuels can be produced from biomass through a process called pyrolysis. Pyrolysis occurs when biomass is heated in the absence of oxygen. The biomass then turns into a liquid called pyrolysis oil, which can be burned like petroleum to generate electricity.

Bioproducts

Biomass can be used to produce a variety of biodegradable plastic products. These bioproducts, or biobased products, are not only made from renewable sources, they also often require less energy to produce than petroleum-based products. Researchers have discovered that the process for making biofuels - releasing the sugars that make up starch and cellulose in plants - also can be used to make antifreeze, plastics, glues, artificial sweeteners, and gel for toothpaste.

SOURCES OF BIOMASS

The term 'biomass' is used to describe any kind of organic matter located within the layer of living systems around the planet, commonly known as the 'biosphere'. Sources of biomass **(Fig 9.2)** include agricultural crops, animal and plant wastes, algae, wood and organic residential/ industrial waste. The type of biomass will determine the type and amount of bioenergy that can be produced and the technology that can be used to produce it.

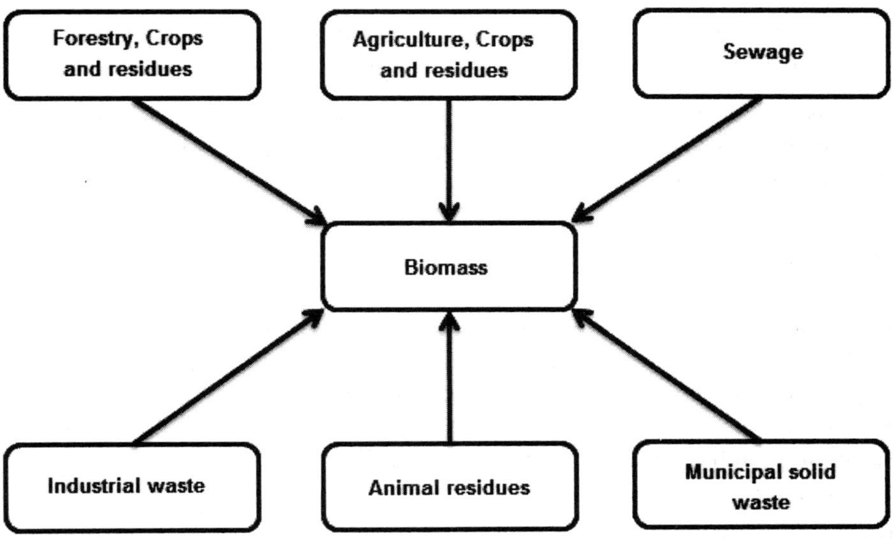

Fig. 9.2: Sources of Biomass

For example, agricultural crops, like corn and canola, can be used to produce liquid biofuels such as ethanol and biodiesel. Alternatively, wet wastes like manure are well suited to produce biogas through anaerobic digestion, which can be combusted to generate electricity and heat or upgraded into a transport fuel, biomethane.

The most obvious source of biomass is trees. Wood fuel can be derived from conventional forestry practice such as thinning and trimming as part of sustainable management of woodland to ensure the production of high quality timber for construction, joinery and wood products, and optimization of biodiversity. It can also be derived from tree surgery operations and the management of parks, gardens and transport corridors. There are many different forms of biomass fuel obtainable from trees, but they are all categorized as virgin wood provided they have not been used for another purpose. Virgin wood consists of wood and other products such as bark and sawdust which have had no chemical treatments or finishes applied. Wood may be obtained from a number of sources which may influence it's physical and chemical characteristics.

Waste is the spoilage, loss or destruction of either matter or energy, which is unusable to man. Gradually increasing civilization through industrialization and urbanization, has led to increase in generation of wastes into environment from various sources. Based on the chemical nature, material wastes are of various types:

a) Inorganic wastes (those generated by metallurgical and chemical industries, coal mines, etc.).

b) Organic wastes (agricultural products, dairy and milk products, slaughter houses, sewage, forestry, etc.).

c) Mixed wastes (those discharged from industries dealing with textiles, dyes, cake and gas, plastic, wool, leather, petroleum, etc.).

The inorganic wastes may be recovered by chemical/ mechanical treatment, whereas organic and mixed wastes require biological as well chemical treatments.

Bioethanol

Bioethanol or simply 'ethanol' is a renewable energy source made by fermenting the sugar and starch components of plant by-products - mainly sugarcane and crops like grain, using yeast. It is also made from corn, potatoes, milk, rice, beetroot and recently grapes, banana and dates depending on the countries agricultural strength.

Today, bioethanol has many uses: It is blended with petrol to make a truly sustainable transport fuel; it's used in cosmetic and other manufacturing processes. Bioethanol is completely composed of biological products. The combustion of bio ethanol results in a clean emission: Heat, Steam and Carbon Dioxide. Carbon dioxide is absorbed by plants. It is then processed via photosynthesis to help the plant grow. This infinite cycle of creation and combustion of energy makes bio ethanol a carbon neutral fuel source. Bioethanol is considered an alternative to petroleum and diesel and its popularity is emerging as a fuel for cars – it is particularly well established in Brazil. Production of bioethanol involves the conversion of a feedstock crop into fermentable sugars through enzyme amylases. Yeast is then added to ferment the sugars into alcohol and carbon dioxide. The main crop used in bioethanol production varies throughout the world – in Brazil, sugar cane is preferred, in the USA its corn and across Europe it's predominantly wheat and barley.

Production of Bioethanol

Almost any plants that composed largely of sugars (e.g. leaves, grains, wheat) can be used to produce ethanol (as sugar is essential in the production of ethanol). During ethanol fermentation, glucose and other sugars in the corn (or sugarcane or other crops) are converted into ethanol and carbon dioxide.

$$\mathbf{C_6H_{12}O_6 \rightarrow 2\ C_2H_5OH + 2\ CO_2 + heat}$$

Starch and cellulose are molecules that are strings of glucose molecules. It is also possible to generate ethanol out of cellulosic materials. However, a pretreatment is necessary that splits the cellulose into glucose molecules and other sugars which subsequently can be fermented. The resulting product is called cellulosic ethanol, indicating its source.

Bioethanol from corn

Corn ethanol is ethanol produced from corn that is used as a biomass. Corn ethanol is produced by means of ethanol fermentation and distillation. There are two main types of corn ethanol production: dry milling and wet milling. The products of each type are utilized in different ways.

Dry milling

In the dry milling process the entire corn kernel is ground into flour and referred to as "meal." The meal is then slurried by adding water. Enzymes are added to the mash that converts starch to dextrose, a simple sugar.

Wet milling

The process of wet milling takes the corn grain and steeps it in a dilute combination of sulfuric acid and water for 24 to 48 hours in order to separate the grain into many components. The slurry mix then goes through a series of grinders to separate out the corn germ. Corn oil is a by-product of this process and is extracted and sold. The remaining components of fiber, gluten and starch are segregated out using screen and centrifugal separators. The corn starch and remaining water can then be processed one of three ways:

(i) Fermented into ethanol, through a similar process as dry milling.

(ii) Dried and sold as modified corn starch.

(iii) Made into corn syrup.

Bioethanol from lignocellulosic materials

Biofuel produced from lignocellulosic materials, so-called second generation bioethanol shows energetic, economic and environmental advantages in comparison to bioethanol from starch or sugar. However, physical and chemical barriers caused by the close association of the main components of lignocellulosic biomass, hinder the hydrolysis of cellulose and hemicellulose to fermentable sugars. The main goal of pretreatment is to increase the enzyme accessibility improving digestibility of cellulose. Each pretreatment has a specific effect on the cellulose, hemicellulose and lignin fraction thus; different pretreatment methods and conditions should be chosen according to the process configuration selected for the subsequent hydrolysis and fermentation steps **(Fig 9.3)**. Biomass must be pre-treated with acids so as to minimize the size of the feedstock before obtaining the sugars in it by the processes of hydrolysis and sugar fermentation. During the hydrolysis process, dilute acids broke down the cellulose and hemi-cellulose fragments into sucrose sugar which then undergo the sugar fermentation process to produce ethanol. There are three main methods of hydrolysis to extract sugars out from biomass:

(i) Concentrated acid hydrolysis

(ii) Enzymatic hydrolysis

(iii) Dilute acid hydrolysis

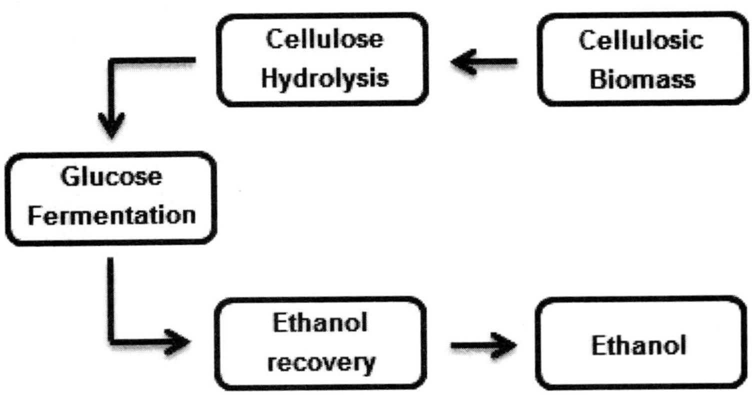

Fig. 9.3: Process of Bioethanol production from cellulosic material

Fermentation
The equipment and processing technology for producing ethanol from cellulose is the same as for producing ethanol from grain. In addition, yeast used in grain-based ethanol production can use glucose obtained from cellulose. However, only about 50-60% of the sugar derived from cellulose-rich plant materials is glucose. The remaining 40-50% is largely a sugar called "xylose," which naturally occurring yeast cannot ferment to ethanol. Fermentation is started by adding yeast to the sugar mixture. The sugars will be broken down to ethanol (a form of alcohol) and carbon dioxide.

Biotechnology has been used to genetically modify yeast and some bacteria to allow them to produce ethanol from both glucose and xylose. These advances increase the amount of ethanol than can be produced from a ton of cellulosic material by as much as 50%. Additional improvements, based upon understanding the basic metabolism and genetics of microorganisms, are underway to increase the efficiency and rates that the microorganisms convert xylose to ethanol.

Distillation
For the ethanol to be usable as a fuel, water must be removed. Most of the water is removed by distillation. The purity is limited to 95-96% due to the formation of a low-boiling water-ethanol azeotrope. This may be used as fuel alone but unlike anhydrous ethanol it is immiscible in Petrol meaning it cannot be mixed i.e. E85. The water fraction is typically removed in further treatment in order to burn with in combination with petrol in petrol engines.

The United States is the world's largest producer of ethanol, having produced over 13 billion gallons in 2012 alone. Together, the U.S. and Brazil produce 87% of the world's ethanol. The vast majority of U.S. ethanol is produced from corn, while Brazil primarily uses sugar.

Biodiesel

Biodiesel is a domestically produced, renewable fuel that can be manufactured from vegetable oils, animal fats, or recycled restaurant grease for use in diesel vehicles. Biodiesel's physical properties are similar to those of petroleum diesel, but it is a cleaner-burning alternative. Using biodiesel in place of petroleum diesel reduces emissions. It consists of long-chain alkyl (methyl, ethyl, or propyl) esters. Biodiesel can be used alone, or blended with petro diesel.

Blends of biodiesel and conventional hydrocarbon-based diesel are products most commonly distributed for use in the retail diesel fuel marketplace. Much of the world uses a system known as the "B" factor to state the amount of biodiesel in any fuel mix 100% biodiesel is referred to as B100, while 20% biodiesel, 80% petro diesel is labeled B20.

The cold-flow properties of biodiesel blends vary depending on the amount of biodiesel in the blend. The smaller the percentage of biodiesel in the blend, the better it performs in cold temperatures. In the transport sector, it may be effectively used both when blended with fossil diesel fuel and in pure form. Tests undertaken by motor manufacturers in the European Union on blends with diesel oil up to 5-10%, or at 25-30% and 100% pure have resulted in guarantees for each type of use.

Benefits

Biodiesel has been demonstrated to have significant environmental benefits in terms of decreased global warming impacts, reduced emissions, greater energy independence and a positive impact on agriculture. Biodiesel is extremely low in sulphur, and has a high lubricity and fast biodegradability.

Biodiesel - Indian Scenario

Biodiesel is now being produced locally in India for use in stationary engines and large or slow engines like those in trains, trucks and tractors. Efforts are also on to use ethanol as a substitute for petrol. Biodiesel is rapidly replacing both kerosene (which was used illegally and inefficiently) and diesel as a more efficient, cheap, and clean alternative for large engines. Biodiesel-blends are being used to run state transport corporation buses in Karnataka. The University of Agriculture Sciences at Bangalore has identified many elite lines of Jatropha curcas and Millettia pinnata (Pongamia tree). Indian Oil Corporation has tied up with Indian Railways to introduce biodiesel crops over 1 million square kilometers. Also, Jharkhand and Madhya Pradesh have tied up with IndianOil to cultivate large tracts of land with Jatropha, the former crop of choice for Indian biodiesel plans. Jatropha is now being replaced by Pongamia and Castor due to its very (comparative) high cost of cultivation. In order to organize the industry, the Biodiesel Society of India has been formed to encourage energy plantations for increasing feedstock supplies.

Bioethanol and Biodiesel Comparison

The two most widely used types of biofuels are ethanol and biodiesel. Just a brief review, ethanol is an alcohol fuel derived from sugarcane, wheat, corn and biomass, thus including wasted cooking oil. It can be blended with conventional petroleum diesel to improve its octane level resulting in reduced greenhouse gas emissions. On the other hand, biodiesel is made from natural oils such as animal fats or vegetable oils. Presently, experts states that the main biofuels, ethanol and biodiesel, have a positive effect to the environment. Each of them has its pros and cons. The utilization of both ethanol and biodiesel will result to a decrease reliance on foreign conventional fossil fuels and a reduction on harmful, toxic emissions. There are various methods to compare ethanol and biodiesel. Below are some of them.

Process of Manufacture

The two processes that can generate fuel-grade ethanol are wet milling and dry milling. But the most widely used is the dry-mill method. In its simplest explanation, yeast, sugars and starch are fermented. From starch, it is fermented into sugar; afterwards it is fermented again into alcohol.

Biodiesel is produced through a chemical process termed as transesterification. In this method, the two byproducts, methyl esters and glycerin which are not good for engines, are left behind.

Environmental Benefit

Both biofuels can reduce harmful emissions. Both biodiesel and ethanol could provide significant environmental benefit. The two have a great probability of decreasing the greenhouse gas emissions because of the fact that these biofuels are primarily derived from crops which absorb carbon dioxide. Thus, the balance of carbon dioxide is sustained and maintained in the atmosphere.

Compatibility

Biodiesel can run in any diesel generated engines. However, it is not yet clear if it is the same when using 100% ethanol since it is recommended to be blended with fossil fuel like gasoline. Therefore, it is only compatible with selected gasoline powered automobiles.

Costs

Currently, commercial biodiesel is more expensive than ethanol.

Algae as a Source of Biofuels

Replacing fossil fuels with algae, a renewable resource, to make biodiesel is an exciting possibility. The term "algae" is used to refer to a diverse group of aquatic, estuarine and marine plant organisms, which range in size from microscopic (microalgae) to many meters in length. There are more than 30,000 fresh water and salt water species of algae. Algae can use various water sources ranging from wastewater to brackish water and be grown in small, intensive plots on denuded land. They come in various forms and colors; from tiny protozoa floating in ponds to huge bunches of seaweed inhabiting the ocean. Leafy kelp, grassy moss and fungus growing on rocks are all forms of algae. While algae may still produce some CO_2 when burned, it can sequester CO_2 during growth in a way that fossil-fuel based energy sources obviously can't. Algae is easy to produce and requires less land to do so than many other plant sources commonly used in the making of fuels, making it an attractive candidate for full-scale biodiesel production. In addition, with a composition containing about half lipid oils, algae appear to be a rich resource as a biofuel feedstock. Algae contain a lot of oil, which is extracted by number of ways.

Algae are grown in either open-pond or closed-pond systems, once the algae are harvested, the lipids, or oils, are extracted from the walls of the algae cells.

Extraction Methods

Different extraction methods are discussed below:

Oil Press

There are a few different ways to extract the oil from algae. The oil press is the simplest and most popular method. It's similar to the concept of the olive press. It can extract up to 75 percent of the oil from the algae being pressed.

Hexane solvent method

Basically a two-part process, the hexane solvent method (combined with pressing the algae) extracts up to 95 percent of oil from algae. First, the press squeezes out the oil. Then, leftover algae is mixed with hexane, filtered and cleaned so there's no chemical left in the oil.

Supercritical fluids method

The supercritical fluids method extracts up to 100 percent of the oil from algae. Carbon dioxide acts as the supercritical fluid -- when a substance is pressurized and heated to change its composition into a liquid as well as a gas. At this point, carbon dioxide is mixed with the algae. When they're combined, the carbon dioxide turns the algae completely into oil. The additional equipment and work make this method a less popular option.

Once the oil's extracted, it is refined using fatty acid chains in a process called transesterification **(Fig 9.4)**. In this process, an alcohol and an ester compound are mixed, and through the resulting reaction produce a different type of alcohol and a different type of ester. Here, a catalyst such as sodium hydroxide is mixed in with an alcohol such as methanol. This creates a biodiesel fuel combined with a glycerol. The mixture is refined to remove the glycerol. This process allows the oil that comes from the algae to be altered into biodiesel through a specific chemical reaction. The final product is algae biodiesel fuel.

The process of extracting oil from the algae is universal, but companies producing algae biodiesel are using diverse methods to grow enough algae to produce large amounts of oil.

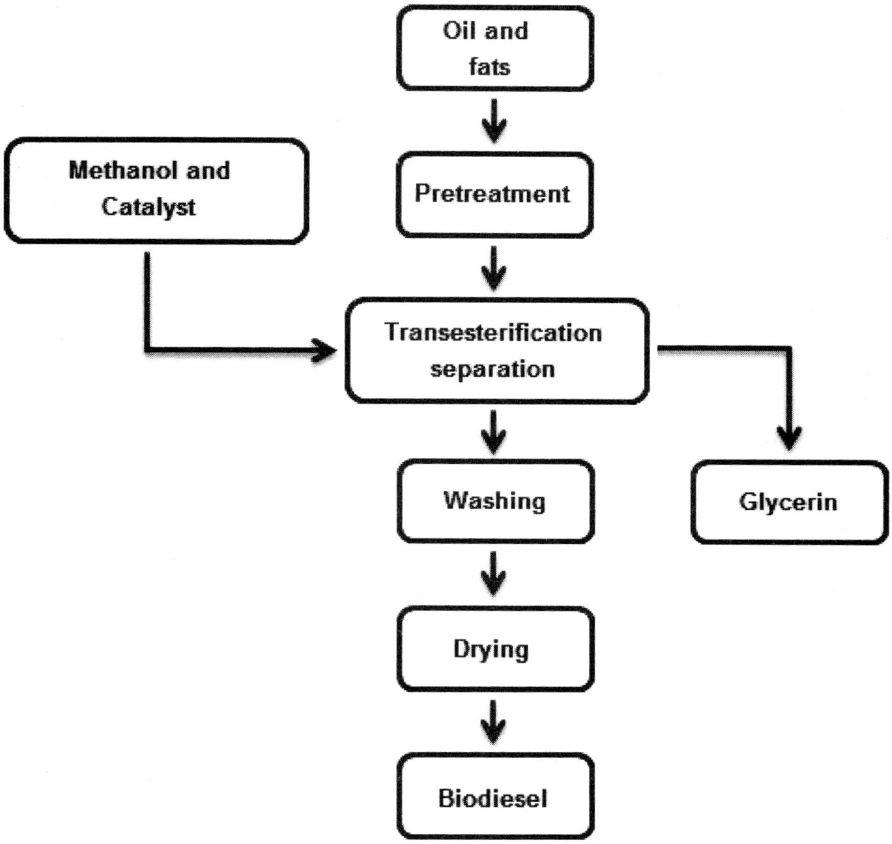

Fig. 9.4: Biodiesel production

Algae Biodiesel Production Methods

Open pond method

The most natural method of growing algae for biodiesel production is through open-pond growing. Using open ponds, algae can be grown in hot, sunny areas of the world to get maximum production. While this is the least invasive of all the growing techniques, it has some drawbacks. Bad weather can stunt algae growth, as can contamination from strains of bacteria or other outside organisms. The water in which the algae grow also has to be kept at a certain temperature, which can be difficult to maintain.

Vertical growth/Closed loop production

Another method for growing algae is a vertical growth or closed loop production system. This process actually came about as biofuel companies sought to produce algae quicker and more efficiently than what was possible utilizing open pond growth. Vertical growing places algae in clear, plastic bags which allow them to be exposed to sunlight on more than just one side. These bags are stacked high and protected from the elements with a cover. While that extra sun may seem trivial, in reality, the clear plastic bag provides just enough exposure to sunlight to increase the rate of the algae production. Obviously, greater the algae production, greater the potential amount of oil that will later be extracted. And unlike the open pond method where algae are exposed to contamination, the vertical growth method isolates algae from this concern.

Closed tank bioreactor

A third method of extraction that biodiesel companies are continuing to perfect is the construction of algae closed-tank bioreactor plants to further increase already-high oil production. In this method, an alga isn't grown outside. Instead, indoor plants are built with large, round drums that are able to grow algae under near perfect conditions. Within these barrels, the algae can be manipulated into growing at maximum levels--even to the point they can be harvested every day. This method, understandably, results in a very high output of algae and oil for biodiesel. Some companies are locating their closed bioreactor plants near energy plants so that extra carbon dioxide can actually be recycled rather than polluting the air.

Researchers are testing another variation of the closed-container or closed-pond process -fermentation. Algae are cultivated in closed containers and fed sugar to promote growth. This method eliminates all margin of error since it allows growers to control all environmental factors. The benefit of this process is that it allows the algae biodiesel to be produced anywhere in the world. But, researchers are trying to figure out where to get enough sugar without creating problems.

Advantages of Algal Biofuels

a) Algae offer some advantages over terrestrial plants for biofuel production. Algal generation times (cell division) and population growth take place over hours and days. Terrestrial plant growth and reproduction typically occur in time frames of months to years.

b) The rapid cell growth and reproduction cycles of many algal species make genetic strain development or modifications easier and potentially much faster. Moreover, using algae for biofuel production would spare corn and soybean for consumption as food by humans.

c) With growing concerns about the accumulation of carbon dioxide in our atmosphere and global warming, some view microalgae biomass production as a potential method for removal or abatement of atmospheric carbon dioxide. The idea is to couple algal biofuel production with coal-fired power plants. Carbon dioxide from the power plant flues would provide the substantial volumes needed by algae for large-scale biomass production.

d) Produces orders of magnitude more oil and hydrocarbons per land mass than any terrestrial crop, and can be cultivated virtually anywhere.

Biogas

Biogas is produced by the anaerobic digestion with anaerobic bacteria or fermentation of biodegradable materials such as manure, sewage, municipal waste, green waste, plant material, and crops. Biogas comprises primarily methane (CH_4) and carbon dioxide (CO_2) and may have small amounts of hydrogen sulphide (H_2S), nitrogen and moisture. In addition to providing electricity and heat, biogas is useful as a vehicle fuel. When processed to purity standards, biogas is called renewable natural gas and can substitute for natural gas as an alternative fuel for natural gas vehicles.

Methods of Production

Biogas is a product of decomposing organic matter, such as sewage, animal byproducts, and agricultural, industrial, and municipal solid waste. Biogas must be upgraded to a purity standard to fuel vehicles and be distributed via the existing natural gas grid.

Anaerobic digestion method

Biomass that is high in moisture content, such as animal manure and food-processing wastes, is suitable for producing biogas using anaerobic digester technology. Anaerobic digestion is a biochemical process in which particular kinds of bacteria digest biomass in an oxygen-free environment. Several different types of bacteria work together to break down complex organic wastes in stages, resulting in the production of "biogas".

Symbiotic groups of bacteria perform different functions at different stages of the digestion process. There are four basic types of microorganisms involved. Hydrolytic bacteria break down complex organic wastes into sugars and amino acids. Fermentative bacteria then convert those products into organic acids. Acidogenic microorganisms convert the acids into hydrogen, carbon dioxide and acetate. Finally, the methanogenic bacteria produce biogas from acetic acid, hydrogen and carbon dioxide.

Available Feedstock for Anaerobic Digestion

Sewage sludge

Digestion of sewage sludge provides significant benefits when recycling the sludge back to land. The digestion process sanitizes and also reduces the odor potential from the sludge. Typically between 30 and 70% of sewage sludge is treated by anaerobic digestion depending on national legislation and priorities. In countries like Sweden and Switzerland limitations for the field application of sludge have been introduced. However, anaerobic digestion is still considered an important step since it produces renewable energy and improves the ability of the sludge to settle which makes it easier to dry. In less developed countries, direct anaerobic digestion is the only treatment of waste water. If the digester is adequately designed and the retention time of the water is long enough, the quality of the treated water can be excellent.

Biogas from Landfills
Methane can be captured from landfills and used to produce biogas. Methane gas collection is practical for landfills at least 40 feet deep with at least 1 million tons of waste.

Biogas from Livestock Operations
Biogas recovery systems at livestock operations can produce renewable energy in cost-effective ways. Animal manure can be collected and delivered to an anaerobic digester to stabilize and optimize methane production. The resulting biogas can be used to fuel natural gas vehicles. The U.S. Environmental Protection Agency (EPA) estimates 8,200 U.S. dairy and swine operations could support biogas recovery systems with the potential to generate more than 13 million megawatt-hours and displace about 1,670 megawatts of fossil fuel-fired generation collectively per year. Biogas recovery systems are also feasible at some poultry operations.

Industrial Wastes
Organic solid wastes from industry are increasingly treated in biogas plants. Even if some of the substances might be difficult to digest as a sole substrate, in mixture with manure or sewage sludge they don't pose any problem. The combined digestion of different wastes is called co-digestion. Most of the waste products from the food industry have excellent gas potential and therefore are in demand by plant operators.

Municipal Solid Wastes
Organic wastes from households and municipal authorities provide potential feedstock for anaerobic digestion. The treatment of clean source separated fractions for recycling of both the energy content and the organic matter is the only method in which the cycle can be completely closed.

Biogas Upgrading
There are a number of compounds which have to be removed from biogas whenever they are present. Most often only water vapor and hydrogen sulphide are to be removed except when gas is compressed as vehicle fuel. Then it is recommended that CO_2 is also removed. When the gas is fed to the

grid it has to meet energy standards which usually require 97% methane. The following compounds might be present in biogas:

(i) Water vapor

(ii) Carbon dioxide

(iii) Hydrogen sulphide

(iv) Siloxane

(v) Aromatic compounds

(vi) Air (oxygen, nitrogen)

(vii) Halogenic compounds (chlorides, fluorides)

Benefits of Biogas

Biogas can be an alternative to conventional transportation fuels. The benefits of biogas are similar to the benefits of natural gas. Additional benefits include:

(i) Increased Energy Security - Biogas offsets non-renewable resources, such as coal, oil, and fossil fuel-derived natural gas.

(ii) Fewer Emissions - Biogas reduces emissions by preventing methane release in the atmosphere. Methane is many times stronger than carbon dioxide as a greenhouse gas.

(iii) Better Economics - Biogas reduces the cost of complying with EPA combustion requirements for landfill gas.

(iv) Cleaner Environment - producing biogas through anaerobic digestion reduces landfill waste and odors, produces nutrient-rich liquid fertilizer, and requires less land than aerobic composting.

(v) Research and development efforts are reducing the costs of biogas production and purification, producing higher-quality natural gas from biogas, and evaluating the performance of biogas-fueled vehicles.

Future Prospects

As with all forms of bioenergy, the future looks bright for biogas technology. As a CO_2 -neutral source of energy it will be increasingly used to meet the Kyoto Protocol commitments and to benefit from the CO_2-emission trade. Biogas is a flexible form of renewable energy that can produce heat, electricity and serve as a vehicle fuel. As well as energy, the anaerobic digestion process yields valuable fertilizer and reduces emissions and odor nuisances. It therefore can make a positive contribution to multiple goals in government programmes.

BENEFITS OF BIO-ENERGY

As concerns grow over both climate change and the fast depleting reserves of fossil fuels, Bio-Energy has a prominent role to play in the provision of clean burning fuels. An added advantage of Bio-Energy is that it also encourages the planting of forests as carbon sinks.

Economic Benefits

(i) Cost competitive fuels - biomass offers a cost-effective alternative to burning fossil fuels for some users. For example - In saw milling, the users can burn the wood waste produced on site.

(ii) Waste minimization - Using agro-waste, organic forest residue, etc., as fuel converts waste to a value proposition.

(iii) Rural Development - Biomass is a local resource; it is produced, processed and consumed locally. It, therefore, emphasizes on the self sufficiency model and encourages sustainable development in the villages thereby generating more employment and income for the villagers and contributing to nation building.

(iv) Energy storage - Unlike wind, wave and solar, biomass is a storable form of renewable energy. It is capable of being transported and utilized at any time.

Environmental Benefits

(i) Carbon neutral - As a renewable energy source that can be grown and used sustainably, burning biomass has zero net greenhouse effect as carbon dioxide given off during combustion is absorbed by the growth of the next crop of biomass.

(ii) Renewable energy source - Unlike fossil fuels, biomass fuels are renewable and therefore contribute to a more sustainable, clean and green future for human beings.

(iii) Across the globe, hundreds of millions of acres of once-productive agricultural land lie abandoned, according to a new report from researchers at Stanford University and the Carnegie Institution for Science. This land can be used to grow crops for conversion into Bioenergy; it could help ease the energy crunch without worsening the world food shortage or contributing to global warming.

FUTURE OF BIO-ENERGY

Many influential organizations foresee biomass playing a key role in a future, more sustainable, global energy supply matrix. Both developed or developing countries are actively encouraging the use of biomass for energy, and pushing forward the development of the necessary knowledge and technology for modern biomass energy systems. There is a growing consensus that renewable energy must progressively displace the use of fossil fuels, with fears of global climate change adding urgency to this need. Among the available types of renewable energy, biomass is unique in its ability to provide solid, liquid and gaseous fuels which can be stored and transported. The potential resource for Bio-Energy is large, especially in forest-rich nations, in richer countries where there is a surplus of available agricultural land and in many low latitude countries where high biomass yields are possible. Therefore we expect biomass to be an important fuel of the future demonstrating clear environmental and social benefits.

BIOSENSORS

A biosensor is an analytical device consisting of an immobilized layer of biological material (e.g. enzyme, antibody, organelle, hormones, nucleic acids or whole cells) in the intimate contact with a transducer i.e. sensor (a physical component) which analyses the biological signals and converts into an electrical signal (Gronow, 1984). A sensor can be anything, a single carbon electrode, an ion-sensitive electrode, oxygen-electrode, or a photocell. In other words every biosensor consists of three parts **(Fig 9.5)**:

a) The "sensitive biological element" (biological material (eg. tissue, micro-organisms, organelles, cell receptors, enzymes, antibodies, nucleic acids, etc), a biologically derived material or biomimic) the sensitive elements can be created by biological engineering.

b) The "transducer" or the "detector element" (works in a physicochemical way; optical, piezoelectric, electrochemical, etc.) that transforms the signal resulting from the interaction of the analyte with the biological element into another signal (i.e., transducers) that can be more easily measured and quantified.

c) Associated electronics or signal processors that are primarily responsible for the display of the results in a user-friendly way.

In 1987, for the first time Yellow Springs Instruments Co., USA developed a biosensor for diagnostic purposes for measuring glucose in blood plasma. It is a hand - held machine which measures six components of blood plasma for example, glucose, urea, nitrogen, sodium, potassium and chloride. Blood glucose biosensor uses the enzyme glucose oxidase to break blood glucose down. In doing so it first oxidizes glucose and uses two electrons to reduce the FAD (a component of the enzyme) to FADH2. This in turn is oxidized by the electrode (accepting two electrons from the electrode) in a number of steps. The resulting current is a measure of the concentration of glucose. In this case, the electrode is the transducer and the enzyme is the biologically active component.

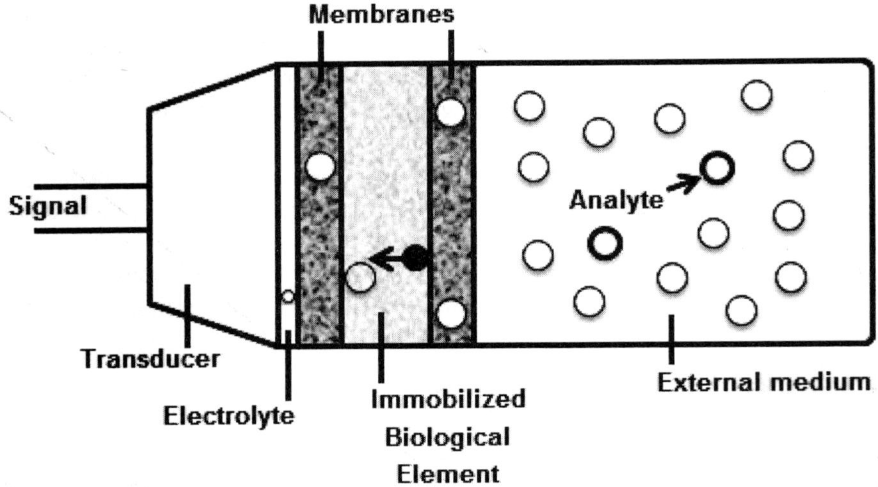

Fig. 9.5: Components of Biosensor

Types of Biosensors

Biosensors are of different types based on the use of different biological material and sensor devices; a few of them are discussed below:

Electro-chemical Biosensor

An electrochemical biosensor exploits the detection of physicochemical properties of electroactive substances to realize the biorecognition that provides the measurable signal: electrical current, voltage, resistance or superficial charge. Electrochemical biosensors are normally based on enzymatic catalysis of a reaction that produces or consumes electrons (such enzymes are rightly called redox enzymes).The sensor substrate usually contains three electrodes; a reference electrode, an active electrode and a sink electrode. An auxiliary electrode (also known as a counter electrode) may also be present as an ion source. The target analyte is involved in the reaction that takes place on the active electrode surface, and the ions produced create a potential which is subtracted from that of the reference electrode to give a signal.

Thermistor Containing Biosensor

Thermistor is used to record even a small temperature changes (between 0.1-0.001°C) during biochemical reactions. Different thermistors have been developed by immobilizing enzymes like cholesterol oxidase, glucose oxidase, invertase, tyrosinase, etc. Moreover, thermistors are also employed for the study of antigen- antibody with very high sensitivity (10-13 mol dm-3) in case of thermometric ELISA.

Bioaffinity Biosensor

Bioaffinity sensors are developed recently. It measures the concentration of the determinants, i.e. substrates based on equilibrium binding. This shows a high degree of selectivity. These are of diverse nature because of the use of radiolabelled, enzyme labeled or fluorescence-labeled substance (Kumar and Kumar, 1992).

Amperometric Biosensor

Amperometric biosensors are those which measure the reaction of analyte with enzyme and generate electrons directly or through a mediator. The Amperometric biosensors contain either enzyme-electrode or without a mediator, or chemically modified electrodes.

Bioluminescent Biosensor

A bioluminescent biosensor exploits the phenomena of visible light emission in biological entities due to the oxidation of organic compounds mediated by a catalytic enzyme. The light generation depends on the chemical reaction kinetics. This type of biosensor requires a photodetector to transduce the light emitted by the biological entity to an electrical signal.

Optical Biosensor

An optical biosensor exploits two different strategies:

(i) Changes in light absorption between the reactants and products of a chemical reaction mediated by enzymes

(ii) Measurements of the output light by a luminescent process.

Piezoelectric Biosensor

A piezoelectric biosensor makes use of the change in frequency of a piezoelectric crystal, which is proportional to the mass of absorbed material as product of the chemical reaction catalyzed by enzymes. A special case is associated with a resonant mirror (RM) biosensor that uses the evanescent field emitted in a waveguide to determine changes in the refractive index at the sensing biochemical surface.

APPLICATIONS OF BIOSENSORS

Biosensors have potential scope of applications in the fields of health care, industrial process control, military operations, environmental monitoring and in food industry. Some of them are as follows:

a) Biosensors are used in detection of pollutants.

b) Biosensors are used to measure the concentration of various metal ions.

c) They are used in detection of BOD during waste water treatment.

d) Biosensors are used in monitoring of glucose.

e) Tumor cells are used as biosensors to monitor the susceptibility of chemotherapeutic drugs.

f) Biosensors are used in organ replacement procedures such as artificial pancreas in diabetic patients.

g) They are also used in biotechnological processes such as to determine proteins or peptides.

h) Biosensors are used to monitor manufacturing of pharmaceutical compounds.

REFERENCES

Abdi, N., Hamadache, F., Belhocine, D., et al. 2000. Enzymatic saccarification of solid residue of olive mill in a batch reactor. Biochem. Eng. J., 6, 77-183.

Aden, A., et al. 2002. Lignocellulosic biomass to ethanol processing design and economics utilizing co-current dilute acid prehydrolysis and enzymatic hydrolysis for corn stover. National Renewable Energy Laboratory Technical Report NREL/TP-510-32438 (http,//www.nrel.gov/docs/fy02osti/32438.pdf).250

Agarfal, A. K. 2007. Biofuels (alcohols and biodiesel) applications as fuels for internel combustion engines. Progress in Energy and Combustion Science, 33(3), 233-271.

Al-Hasan, M. I. & Al-Momany, M. 2008. The effect of iso-butanol-diesel blends on engine performance. Transport, 23(4), 306-310.

Boychenko, S., Shkilnuk, I., Turchak, V. 2008. The problems of biopollution with jet fuels and the way of achieving solution. Transport, 23(3), 253-257.

Brandon, S. K., Eiteman, M. A., Patel, K., at al. 2008. Hydrolysis of Tifton 85 bermudagrass in a pressurizead batch hot water reactor. J. Technol. Biotechnol., 83, 505-512.

Bremers, G., Birzietis, G., Blija, A., Škele1, A., Danilevics, A. 2009. Bioethanol congruent dehydratation. In, Proceedings of the 8th International Scientific Conference 'Engineering for Rural Development'. Latvia University of Agriculture, Jelgava, 376, 148–155.

Butkus, A., Pukalskas, S., Bogdanovicius, Z. 2007. The influence of turpentine additive on the ecological parameters of diesel engines, Transport, 22(2), 80-82.

Carillo, F., Lis, M. J., Colom, X., et al. 2005. Effect of alcali pretreatment on cellulose hydrolysis of wheat straw, Kinetic study., 40, 3360-3364.

Chandra, R. P., Bura, R., Mabee, W. E., et al. 2007. Substrate pretreatment, the key to effective enzymatic hydrolysis of lignocellulosics. Adv. Biochem. Engin/ Biotechnol., 108, 67-93.

Demirbas, A. 2006. Global biofuel strategies. Energy Edu. Sci. Technol., 17, 27-63.

Demirbas, A. 2009. Biohydrogen, For Future Engine Fuel Demands. In Chapter 3. Biofuels, pp. 61-84. Springer Verlag.

Fuel Ethanol Specifications Brazil. 2008-2010.National Department of Fuels Technical Regulation DNC - 01/91. Specifications for Anhydrous Fuel Ethanol ("AEAC") and Hydrous Fuel Ethanol "AEHC").

Goldstein, I. R., Pereira, H., Pittman, L. J., et al. 1983. The hydrolysis of cellulose with supercontcentated hydrochloric acid. Biotech. Bioeng. Symp., 13, 17-25.

Narendranath NV, Hynes SH, Thomas KC, Ingledew WM. (1997). Effects of lactobacilli on yeast-catalyzed ethanol fermentations. Applied and Environmental Microbiology 63(11):4158-4163.

Oliveira, RPS, Torres, BR, Silli, M, Marques, DAV, Basso, LC and Converti, A. (2009). Use of sugar cane vinasse to mitigate aluminum toxicity to Saccharomyces cerevisiae. Archives of Environmental Contamination and Toxicology, v. 57(3): 488-494.

Oura E. (1977). Reaction-Products of Yeast Fermentations. Process Biochemistry 12(3):19.

Parrou, J.L., Teste, M.A. e Francois, J. (1997). Effects of various types of stress on the metabolism of reserve carbohydrates in Saccharomyces cerevisiae: genetic evidence for a stress-induced recycling of glycogen and trehalose. Microbiology, 143:1891-1900.

Pilgrim, C. Status of the worldwide fuel alcohol industry. (2009) In: The Alcohol Text Book (5th edition), Ingledew,WM, Kelsall, DR, Austin, GD and Kluhspies, C (ed.), Nottingham University Press, Nottingham, 541p.

Pimentel D (2003). Ethanol fuels: Energy balance, economics, and environmental impacts are negative. Natural Resources Research. No. 12, Vol. 2. pp. 127-134.

RFA (2011). Renewable Fuels Association. In: Statistics. Access Feb 17th 2011. Available from: http://www.ethanolrfa.org/pages/statistics.

Scheller, F.W., Schubert, F., Renneberg, R. & Mes82,5~ller, H-G. (1985). Biosensors: Trends and commercialization. Biosensors 1, 135-160.

Turner, A.P.F. (1987). Biosensors: Principles and potential. In Chemical aspects of food enzymes, ed. A.T.Andrews, pp 259-270, London: Royal Society of Chemistry.

Turner, A.P.F., Karube, I. & Wilson, G.S. eds. (1987). Biosensors: Fundamentals and applications, Oxford, U.K: Oxford University Press.

Vadgama, P. (1986). Urea pH electrodes: Characterization and optimization for plasma measurements. Analyst 111, 875-878.bp.

10

Biomining and Bioleaching

Mining is the extraction of valuable minerals or other geological materials from the earth from an ore body, lode, vein, seam, or reef, which forms the mineralized package of economic interest to the miner. Ores recovered by mining include metals, coal and oil shale, gemstones, limestone, and dimension stone, rock salt and potash, gravel, and clay. Mining is required to obtain any material that cannot be grown through agricultural processes, or created artificially in a laboratory or factory. Mining in a wider sense includes extraction of any non-renewable resource such as petroleum, natural gas, or even water. Mining of stone and metal has been done since pre-historic times. Modern mining processes involve prospecting for ore bodies, analysis of the profit potential of a proposed mine, extraction of the desired materials, and final reclamation of the land after the mine is closed.

The nature of mining processes creates a potential negative impact on the environment both during the mining operations and for years after the mine is closed. This impact has led to most of the world's nations adopting regulations to moderate the negative effects of mining operations. Safety has long been a concern as well, and modern practices have improved safety in mines significantly.

MINING PROCESS

The process of mining from discovery of an ore body through extraction of minerals and finally to returning the land to its natural state consists of several distinct steps. The first is discovery of the ore body, which is carried out through prospecting or exploration to find and then define the extent, location and value of the ore body. This leads to mathematical resource estimation to estimate the size and grade of the deposit.

This estimation is used to conduct a pre-feasibility study to determine the theoretical economics of the ore deposit. This identifies, early on, whether further investment in estimation and engineering studies is warranted and identifies key risks and areas for further work. The next step is to conduct a feasibility study to evaluate the financial viability, technical and financial risks and robustness of the project.

Once the analysis determines a given ore body is worth recovering, development begins to create access to the ore body. The mine buildings and processing plants are built and any necessary equipment is obtained. The operation of the mine to recover the ore begins and continues as long as the company operating the mine finds it economical to do so. Once all the ore that the mine can produce profitably is recovered, reclamation begins to make the land used by the mine suitable for future use.

Mining Techniques

Mining techniques can be divided into two common excavation types: surface mining and sub-surface (underground) mining. Today, surface mining is much more common, and produces, for example, 85% of minerals (excluding petroleum and natural gas) in the United States, including 98% of metallic ores.

Surface Mining

It is done by removing (stripping) surface vegetation, dirt, and if necessary, layers of bedrock in order to reach buried ore deposits. Techniques of surface mining include; Open-pit mining which consists of recovery of materials from an open pit in the ground, quarrying or gathering building materials from an open pit mine, strip mining which consists of stripping surface layers off to reveal ore/seams underneath, and mountaintop removal, commonly associated with coal mining, which involves taking the top of a mountain off to reach ore deposits at depth. Most (but not all) placer deposits, because of their shallowly buried nature, are mined by surface methods. Landfill mining, finally, involves sites where landfills are excavated and processed.

Sub-Surface Mining

This mining consists of digging tunnels or shafts into the earth to reach buried ore deposits. Ore, for processing, and waste rock, for disposal, are brought to the surface through the tunnels and shafts. Sub-surface mining can be classified by the type of access shafts used, the extraction method or the technique used to reach the mineral deposit. Drift mining utilizes horizontal access tunnels, slope mining uses diagonally sloping access shafts and shaft mining consists of vertical access shafts. Mining in hard and soft rock formations require different techniques.

ENVIRONMENTAL IMPACT OF MINING

The environmental impact of mining includes erosion, formation of sinkholes, loss of biodiversity, and contamination of soil, groundwater and surface water by chemicals from mining processes. In some cases, additional forest logging is done in the vicinity of mines to increase the available room for the storage of the created debris and soil. Besides creating environmental damage, the contamination resulting from leakage of chemicals also affects the health of the local population. Erosion of exposed hillsides, mine dumps, tailings dams and resultant siltation of drainages, creeks and rivers can significantly impact the surrounding areas, a prime example being the giant Ok Tedi Mine in Papua New Guinea. In areas of wilderness mining may cause destruction and disturbance of ecosystems and habitats and in areas of farming it may disturb or destroy productive grazing and croplands. In urbanized environments mining may produce noise pollution, dust pollution and visual pollution.

The environmental consequences of mining and quarrying are vast, due to an increase in consumerism, heightened dependence on technology, and an overall world population growth, more and more minerals and metals need to be extracted from the earth's surface. Even though the recycling industry is growing, it cannot meet these growing demands for minerals. Furthermore, with numerous accessible mining sites already exploited, the human race is in a constant struggle to find feasible mining methods. One way to mitigate the environmental consequences of mining and quarrying is through the use of bioleaching microbes. These "rock-munching microbes" such as the Acidithiobacillus and Leptospirillum bacterium, can clean up abandon mine sites.

The microbes can detoxify the material present in ores and accelerate the breakdown of minerals. These bacteria can be used in two ways:

Clearing Toxic Waste from Mining Sites

For example, some bacteria have the capacity to stabilize uranium contaminated sites. If they are used properly, they could reduce the chances of such sites contaminating major waterways and ecosystems.

Sustainable Mining

Bioleaching and bio-oxidation processes are alternatives to conventional smelting processes, which discharge large amounts of carbon dioxide, sulfur dioxide, and various toxic materials such as arsenic.

BIOMINING AND BIOLEACHING

Biomining is defined as extracting mineral ores or enhancing the mineral recovery from mines using micro-organisms instead of traditional mining methods. Biomining is an application of biotechnology in recovery of various mineral from ore. Biomining is becoming popular because it is cheap, reliable, efficient, safe, and environmentally friendly, unlike traditional mining methods.

The efficiency of biomining can be increase either by finding suitable strains of microorganisms or by genetically modifying existing microorganisms, made possible due to rapid advances in the field of biotechnology and microbiology. Ores of high quality are rapidly being depleted and biomining allows environment friendly ways of extracting metals from low grade ores.

Biomining includes two different chemical processes called bio-leaching and bio-oxidation. Thus, biomining is an application of biotechnology and is also known as microbial leaching or alternately, biooxidation.

The potential applications of biotechnology to mining and processing are countless. Some examples of past projects in biotechnology include a biologically assisted in situ mining program, biodegradation methods, passive bioremediation of acid rock drainage, and bioleaching of ores and concentrates. This research often results in technology implementation for greater efficiency and productivity or novel solutions to complex problems. Additional capabilities include the bioleaching of metals from sulfide materials, phosphate ore bioprocessing and the bioconcentration of metals from solutions. One project recently under investigation is the use of biological methods for the reduction of sulfur in coal-cleaning applications. From in situ mining to mineral processing and treatment technology, biotechnology provides innovative and cost-effective industry solutions.

Bioleaching is the extraction of metals from their ores through the use of living organisms. This is much cleaner than the traditional heap leaching using cyanide. Bioleaching is one of several applications within biohydrometallurgy and several methods are used to recover copper, zinc, lead, arsenic, antimony, nickel, molybdenum, gold, silver, and cobalt.

The overall advantages of integrating bio-leaching into mining strategies, in addition to sustainability and maintenance are listed below:

(i) Ores of lower metal concentration can be treated economically; this is not feasible using traditional methods. Difficult refractory concentrates can also be processed. This method is emerging as an increasingly important way to extract valuable minerals when conventional methods such as smelting are too expensive.

(ii) Concentrates with contaminants like arsenic, bismuth and magnesia are often expensive to treat in conventional metal-production. Mining companies often have to pay penalties for these hard to treat contaminants when they sell the concentrate to smelters. Using bioleaching microbes can help avoid these large penalties by removing arsenic and other hazardous materials from the concentrates in an environmentally stable form.

(iii) Economic exploitation of smaller deposits, in remote locations, becomes viable because of reduced infrastructural costs.

(iv) Bio-mining allows for the rapid start-up of the mining site, is easy and reliable with regard to maintenance and infrastructural cost and is not labor intensive making it more profitable.

(v) The process takes place at atmospheric pressure and lower temperatures than smelting. Thus, the energy consumption at the mining site is less.

The inputs to the process of bioleaching mainly depend on the type of microorganism used and should be extremely favorable for the dissolution of the metal. Inputs to the process include the following:

a) Metal ore or concentrate to provide energy for bioleaching microbes.

b) Proper air conditions to suit microorganism based on whether they are aerobic or anaerobic.

c) Carbon dioxide because bioleaching microbes need the macro-nutrient carbon to build cell mass.

d) Nitrogen, Phosphorus, Potassium, Magnesium; nutrients needed for bioleaching microbes.

e) PH-regulators, as some microorganisms need highly acidic environments while others need highly alkaline conditions.

f) Bioleaching microbes such as *T. ferrooxidans, T. thiooxidans* & *L. ferrooxidans*.

g) Methods to cultivate bioleaching microbes for inoculation.

h) Temperature control mechanisms.

i) Distribution system, stirring (in tanks), sprinklers, airflow, tubes allowing for the circulation of microbes.

Microorganisms used for Leaching

The most commonly used microorganisms for bioleaching are *Thiobacillus thiooxidans* and *T. ferrooxidans*. The other microorganisms may also be used in bioleaching viz., *Bacillus licheniformis, B. luteus, B. megaterium, B. polymyxa, Leptospirillum ferrooxidans, Pseudomonas fluorescens, Sulfolobus acidocaldarius, Thermothrix thioparus, Thiobacillus thermophilica*, etc.

Thiobacilli

The principal bacteria which play the most important role in solubilizing sulfide metal minerals at moderate temperatures are species of the genus Thiobacillus. They are gram negative rods, either polarly or nonflagellated. Most species are acidotolerant, some even extremely acidotolerant and acidophilic. Some grow best at pH 2 and may grow at pH 1 or even at pH 0.5. Most species are tolerant against heavy metal toxicity. *Thiobacilli* are chemolithoautotrophs that means CO_2 may be the only source of carbon and they derive their energy from a chemical transformation of inorganic matter **(Fig 10.1)**. All *Thiobacilli* oxidize sulfur or sulfur compounds to sulfate or sulfuric acid. If they oxidize hydrogen sulfide, thiosulfate, polythionates or elemental sulfur they produce hydrogen ions and so they lower the pH of the medium, often below pH 2, in some cases below pH 1.

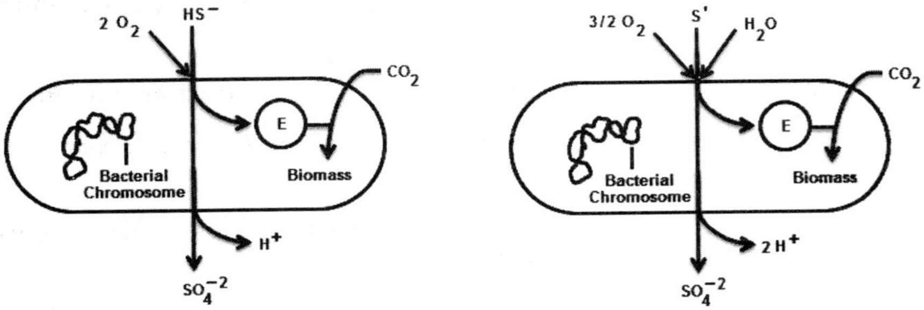

Fig. 10.1: Solubilization of hydrogen Sulfide and elemental sulphur by *Thiobacillus*

Thiobacillus Ferrooxidans

In addition to the oxidation of sulfur and sulfur compounds Thiobacillus ferrooxidans is able to oxidize ferrous to ferric iron and so derive its energy from this exergonic reaction. In this reaction hydrogen ions are consumed and so the pH of the medium should rise **(Fig 10.2)**. But at pH values higher than 2 the ferric iron precipitates as ferric hydroxide, jarosites or similar compounds and this result in the formation of hydrogen ions, so that the pH of the medium is lowered as is the case with oxidation of sulfur compounds.

Thiobacilli and Sulfidic Minerals

Some Thiobacilli, especially *T. ferrooxidans*, are able to oxidize sulfide and some heavy metals -mainly iron but also copper, zinc, molybdenum and presumable some other metals - in the form of sulfidic heavy metal minerals which are of very low solubility in water, practically insoluble. These oxidations result in a solubilization of the minerals.

Other Bacteria

In addition to Thiobacilli there are some other bacteria known to be effective in solubilizing sulfidic minerals. In hot biotopes containing sulfur or oxidisable sulfur compounds, such as hydrothermal vents and self heating brown coal dumps, one can find an archaebacterium named *Sulfolobus*. Like *Thiobacilli* it is acidophilic, chemolithoautotroph and derives its energy from oxidation of sulfur and sulfur compounds and from oxidation of ferrous iron like *Thiobacillus ferrooxidans*. Its pH-range of growth is pH 1.0 - 6.0 and its optimum at about pH 2. A salient characteristic is its thermophily: its growth range is 45 – 85°C, its optimum 70 – 75°C. Species of this genus, especially *S. brierleyi* seem to be the main agent in metal leaching at high temperatures.

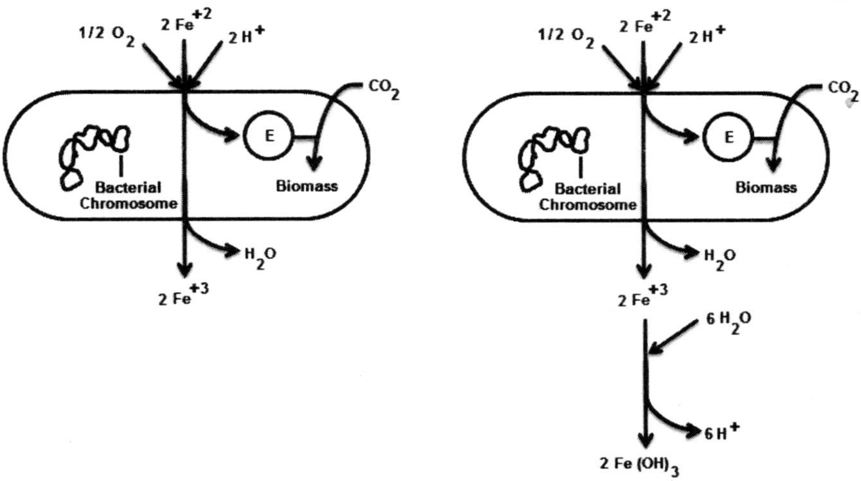

Fig. 10.2: Oxidation of ferrous by Thiobacillus ferrooxidans

Leptospirillum

Often one can see in acid metal leaching biotopes spirilloid bacteria. They belong to the species Leptospirillum ferrooxidans, a gram negative spirillum, facultatively chemolithoautotroph, deriving its energy from oxidizing ferrous iron like Thiobacillus ferrooxidans. But in contrast to this latter bacterium it cannot oxidize sulfur or sulfur compounds and is incapable of utilizing the iron of sulfidic minerals. Leptospirillum ferrooxidans alone cannot solubilize sulfidic ferrous iron containing minerals. But in cooperation with Thiobacillus thiooxidans, which, for its part alone, is also unable to dissolve sulfidic minerals, it can; both bacteria together disintegrate sulfidic ferrous iron containing minerals by oxidation and bringing them into solution.

Chemistry of Microbial Leaching

T. thiooxidans and *T. ferrooxidans* have always been found to be present in mixture on leaching dumps. *Thiobacillus* is the most extensively studied Gram-negative bacillus bacterium which derives energy from oxidation of Fe^{2+} or insoluble sulphur. In bioleaching there are two following reaction mechanisms:

Direct Bacterial Leaching

In direct bacterial leaching a physical contact exists between bacteria and ores and oxidation of minerals takes place through several enzymatically catalyzed steps. For example, pyrite is oxidized to ferric sulphate as below:

$$2FeS_2 + 7O_2 + 2H_2O \xrightarrow{T.\ ferrooxidans} 2FeSO_4 + 2H_2SO_4$$

Indirect Bacterial Leaching

In indirect bacterial leaching microbes are not in direct contact with minerals but leaching agents are produced by microorganisms which oxidize them.

$$FeS_2 + Fe_2(SO_4) \longrightarrow 3FeSO_4 + 2S^\circ$$

$$2S^\circ + 3O_2 + 2H_2O \longrightarrow 2H_2SO_4$$

Oxidation of ferrous (Fe^{2+}) to ferric (Fe^{3+}) by *T. ferrooxidans* at low pH is given below:

$$4FeSO_4 + 2H_2SO_4 + O2 \xrightarrow{T.\ ferrooxidans} 2Fe_2(SO_4)_3 + 2H_2O$$

Types of Biomining

Stirred Tank Biomining

This method is used for leaching from substrates with high mineral concentration. Special types of stirred tank bioreactors lined with rubber or corrosion resistant steel and insulated with cooling pipes or cooling jackets are used for this purpose.

Thiobacillus is commonly used bacteria. Since it is aerobic the bio reactor is provided with an abundant supply of oxygen throughout the process provided by aerators, pumps and blowers. This is multistage process consisting of large no. of bio reactors connected to each other. The substrates moves from one reactor to another and in the final stage it is washed with water and treated with a variety of chemicals to recover the mineral. The name is fairly self-explanatory, as the process requires constructing large aerated tanks that are generally arranged in a series. So that runoff from one tank serves as raw material for the next. In this way, the reactor can operate in continuous flow mode, with fresh ore being added to the first tank while the runoff from the final tank is removed and treated. The ore be processed is generally crushed to a very small particle size to ensure that the solids remain suspended in the liquid medium. Mineral nutrients in the form of $(NH_4)_2SO_4$ and KH_2PO_4 are also added to the tanks to ensure maximal microbial density is maintained.

Due to extremely high cost of stirred tank reactors, they are only used for highly valuable minerals or materials. For gold extraction for example, this technique is usually used when the ore body contains high concentration of arsenopyrite (AsFeS).

Characteristics of mineral degradation in stirred-tank reactors

The environment in a mineral-biooxidation continuous-flow stirred-tank reactor is highly homogeneous as it is operated at a set pH and temperature and with controlled aeration. However, conditions (such as concentrations of soluble metals and metalloids) will vary in a continuous-flow series of tanks as mineral oxidation becomes increasingly extensive, and this can have a significant impact on diversity and numbers of indigenous microbial species. The homogeneity within an individual tank results in a limited ecological niche that tends often to be dominated by two to four species, although smaller numbers of other micro-organisms may be present. Stirred-tank reactors possess three major advantages that are also shared by sewage treatment but by very few other industrial processes. These advantages are linked and together result in the selection of highly efficient microbial consortia.

a) Firstly, the process operates in continuous-flow mode. The advantage derived from continuous-flow operation is that it results in the continual selection for those micro-organisms that are able to grow most efficiently in the tanks. The most efficient growers will be subject to less cell washout and therefore dominate the microbial population.

b) Secondly, the objective of the process is to degrade the substrate (mineral) as rapidly as possible. The mineral provides the energy source and some nutrients for the micro-organisms, and those organisms that are most efficient at degrading the mineral will tend to dominate the process. This means that there will be continual selection for micro-organisms that either catalyze mineral breakdown the fastest or create the conditions in which mineral breakdown occurs most rapidly.

c) Thirdly, process sterility is not required. As the object of the process is the degradation of the mineral, it does not matter which specific organisms carry out this decomposition, and those organisms that do this most efficiently are typically the most desirable. Since the process is non-sterile there is continual selection for micro-organisms that may enter the tanks (e.g. in the concentrate feed) that are more efficient than the resident organisms. This selection includes the selection for genes present in the horizontal gene pool (e.g. genes for metal resistance) that might improve the efficiency of the resident micro-organisms.

Bioheaps

Bioheaps are large amount of low grade ore and effluents from extraction processes that contain trace amounts of minerals. Such effluents are usually stacked in the large open space heaps and treated with microorganisms to extract the minerals. Bioheaps are also called biopiles, biomounds and biocells. They are also used for biodegradation of petroleum and chemical wastes. The low grade ore are crushed and acid treated ore is then agglomerated so that the finer particles get attached to the coarser ones and then treated with the water and effluent liquid. This is done to optimize moisture content in the ore bacteria that is inoculated along with the liquid. The ore is then stacked in the large heaps of 2-10m feet high with aerating tube to provide air supply to the bacteria thus promoting bio oxidation **(Fig 10.3)**.

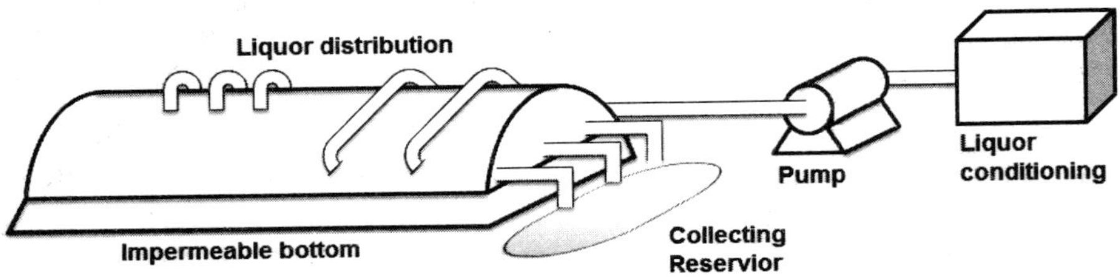

Fig. 10.3: Heap bioleaching

Characteristics and Challenges of Heap Reactors

The engineering design of heaps used to leach ores continues to be refined. Heaps are constructed to pre-determined dimensions using graded ores, irrigated from above with acidic liquors and aerated from below (to provide carbon dioxide required by autotrophic mineral-oxidizing micro-organisms, as well as the oxygen to promote iron- and sulfur-oxidation). However, even the most carefully engineered heap reactors are inevitably heterogeneous (both spatially and temporally), in terms of irrigation efficiency, temperature, pH, the presence of anaerobic pockets, redox potential, dissolved solutes, available nutrients, etc. This lack of homogeneity results in a large number of microenvironments compared with the relatively homogeneous environment provided by a stirred tank. The variability in microenvironment would be expected to support a much greater diversity of mineral-oxidizing and other micro-organisms that colonize different zones and microsites within them. For example, temperatures will be determined by climatic conditions (particularly in the outer layers of a heap), exothermic chemical reactions and heat transfer (conduction, convection and radiation at the heap surface). The oxidation of sulfidic minerals is an exothermic reaction, although heat generation varies between minerals, and is related to their reactivity. Pyrrhotite (FeS) is a more reactive mineral than pyrite (FeS2) and consequently significant heat is often generated in a Pyrrhotite-rich heap, shortly after construction and commissioning. Mineral-oxidizing and other acidophilic prokaryotes often have widely different temperature optima and ranges, and may be conveniently grouped into mesophiles (20–40 °C; predominantly bacteria), moderate thermophiles (40–60 °C; bacteria and archaea) and (extreme) thermophiles (60–80 °C; predominantly archaea). In a heap reactor that experiences fluctuations in temperature, these different groups would be predicted to become more or less dominant, as temperatures increase or decline, assuming that they are present in the first place. Some prokaryotes, notably *Sulfobacillus* spp. And other Firmicutes, are better adapted to survive adverse conditions, such as excessively high or low temperatures, or water stress (zones and microsites within heaps may experience periodic drying, in contrast to stirred tanks) due to their ability to survive as endospores.

It may therefore be predicted that, unlike stirred tanks that are dominated by a small number of indigenous prokaryotes, heap reactors contain a much greater biodiversity, and that the dominant species will vary spatially and during different stages of the life of a heap. There have been relatively few studies on the microbiology of heap bioreactors, and some of these have analysed the liquid phases rather than the ore itself. Most studies have been on chalcocite (Cu_2S) heaps, as this copper mineral is particularly amenable to bioleaching. Microbiological data from analysis of heap populations show that a considerable diversity of acidophiles may be present in these reactors.

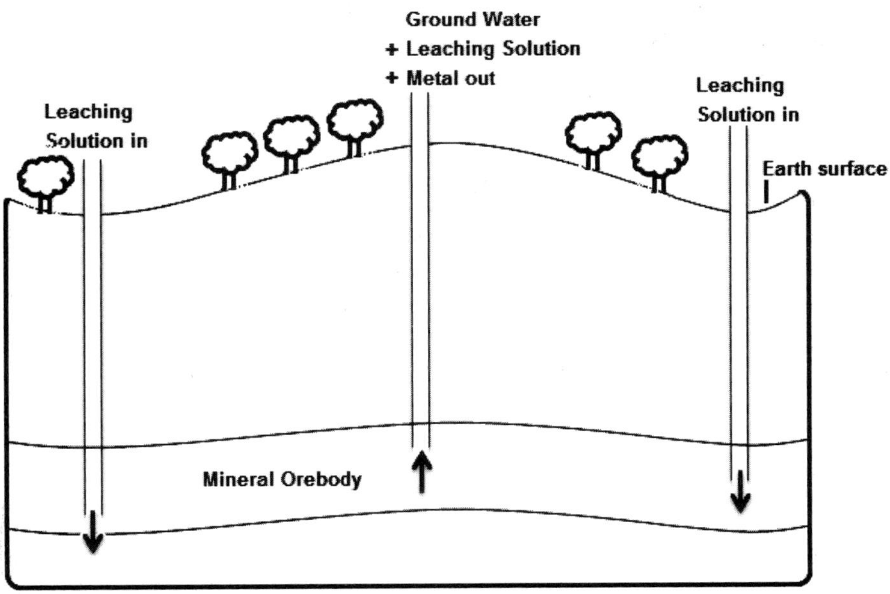

Fig. 10.4: *in situ* Bioleaching

In- situ Bioleaching

In this method the mineral is extracted directly from the mine instead of collecting the ore and transferring to an extracting facility away from the site of the mine. *In situ* Biomining is usually done to extract trace amounts of minerals present in the ores after a conventional extraction process is completed. The mine is blasted to reduce the ore size and to increase permeability and is then treated with water and acid solution with bacterial inoculum. Air supply is provided is provided using pipes or shafts. Biooxidation takes place *in situ* due to growing bacteria and results in the extraction of mineral from the ore. Bio mining promises to transform low grade minerals in to high grade minerals **(Fig 10.4)**. This method is used to improve recovery rates of operation and to reduce operating cost.

Examples of Bioleaching

Copper Leaching

Throughout the world copper leaching plants have been widely used for many years. It is operated as simple heap leaching process or combination of both heap leaching and in situ leaching process. Dilute sulphuric acid (pH 2) is percolated down through the pile. The liquid coming out of the bottom of pile reach in mineral. It is collected and transported to precipitation plant, metal is reprecipitated and purified. Liquid is pumped back to top of pile and cycle is repeated. For removal of copper the ores commonly used are chalcocite (Cu_2S), chalcopyrite ($CuFeS_2$) or covellite (CuS). Several other metals are also associated with these ores. Chalcocite is oxidized to soluble form of copper (Cu^2+) and covellite by *T. ferrooxidans*.

$$\mathbf{Cu_2S + O_2 \longrightarrow CuS + Cu^{2+} + H_2O}$$

Covellite is oxidized to copper sulphate chemically or by bacteria.

$$\mathbf{2CuFeS_2 + 8\tfrac{1}{2}\,O_2 + H_2SO_4 \longrightarrow 2CuSO_4 + Fe_2(SO_4)_3 + H_2O}$$

Thereafter, strictly chemical reaction occurs and is the most important reaction in copper leaching.

$$\mathbf{CuS + 8Fe^{3+} + 4H_2O \longrightarrow Cu^{2+} + SFe^{2+} + SO_4^{2-} + 8H^+}$$

Copper is removed as below:

$$\mathbf{Fe^{2+} + Cu^{2+} \longrightarrow Cu^\circ + Fe^{2+}}$$

Fe^{2+} is transferred in oxidation pond

$$\mathbf{Fe^{2+} + \tfrac{1}{4}O_2 + H^+ \text{-------}> Fe^{3+} + \tfrac{1}{2}O_2}$$

The Fe^{3+} ions produced is an oxidation of ores; therefore, it is pumped back to pile. Sulphuric acid is added to maintain pH. Microbial leaching of copper has been widely used in the USA, Australia, Canada, Mexico, South Africa and Japan. In the USA 200 tonnes of copper is recovered per day.

Uranium Leaching

Uranium leaching is more important than copper, although less amount of uranium is obtained than copper. For getting one tones of uranium, a thousand tons of uranium ore must be handled. In situ uranium leaching is gaining vast acceptance. However, uranium leaching from ore on a large scale is widely practiced in the USA, South Africa, Canada and India. Insoluble tetravalent uranium is oxidized with a hot H_2SO_4/Fe^{3+} solution to make soluble hexavalent uranium sulfate at pH 1.5-3.5 and temperature 35°C.

$$\mathbf{UO_2 + Fe_2(SO_4)_3 \longrightarrow UO_2SO_4 + 2FeSO_4}$$

Uranium leaching is indirect process. *T. ferrooxidans* does not directly attack on uranium ore, but on the iron oxidant. The pyrite reaction is used for the initial production of Fe^{3+} leach solution.

$$\mathbf{2FeS + H_2O + 7\tfrac{1}{2}\, O_2 \longrightarrow Fe2(SO_4)_3 + H_2SO_4}$$

Gold and Silver Leaching

Today's microbial leaching of refractory precious metal ores to enhance gold and silver recovery is one of the most promising applications. Gold is obtained through bioleaching of arsenopyrite/pyrite ore and its cyanidation process. Silver is more readily solubilized than gold during microbial leaching of iron sulfide.

Silica Leaching

Magnesite, bauxite, dolomite and basalt are the ores of silica. Mohanty et al (1990) isolated *Bacillus Ucheniformis* from magnesite ore deposits. Later it was shown to be associated with bioleaching, concomitant mineral lysis and silican uptake by the bacterium. It was concluded that silican uptake was restricted adsorption of bacterial cell surface rather than internal uptake through the membrane. The bioleaching technology of silica magnesite by using *B. licheniformis* developed at Bose Institute, Calcutta is being used for the first time for commissioning a 5 billion tonnes capacity of pilot plant at Salem Works of Burn, Standard Co. Ltd, Tamil Nadu, in collaboration with the Department of Biotechnology, government of India.

The mining industry has increased biological leaching techniques for various reasons, including environmental concerns related to smelting, the decline in the quality of ore reserves, and difficulties in processing. This new interest has motivated increased research. We now know that ore heaps contain a much wider range of organisms than previously thought. In fact, a succession of microbial populations occurs during the leaching of sulfide minerals. Heterotrophic acidophiles belonging to the genera Acidiphilium and Acidocella are found frequently, often in close association with A. ferrooxidans. These heterotrophic species probably scavenge organic molecules that are metabolic byproducts of the chemolithotrophs. Perhaps this association is detrimental, or perhaps it helps A. ferrooxidans thrive by removing wastes. Research continues into the composition of bacterial communities that occur naturally in bioleaching activities. Because ore heaps get quite hot during bioleaching, scientists are also asking whether novel bacteria - perhaps thermophiles from Yellowstone or deep-sea vents - might be seeded onto heaps to provide more efficient biomining.

REFERENCES

Battaglia-Brunet, F., Clarens, M., d'Hugues, P., Godon, J. J., Foucher, S. & Morin, D. (2002). Monitoring of a pyrite-oxidising bacterial population using DNA single strand conformation polymorphism and microscopic techniques. Appl Microbiol Biotechnol 60, 206–211.

Battaglia-Brunet, F., Joulian, C., Garrido, F., Dictor, M.-C., Morin, D.,Coupland, K., Johnson, D. B., Hallberg, K. B. & Baranger, P. (2006).Oxidation of arsenite by Thiomonas strains and characterization of Thiomonas arsenivorans sp. nov. Antonie van Leeuwenhoek 89, 99–108.

Bruhn, D. F., Thompson, D. N. & Naoh, K. S. (1999). Microbial ecology assessment of a mixed copper oxide/sulfide dump leach operation. In Biohydrometallurgy and the Environment. Toward the Mining of the 21st Century, Process Metallurgy 9A, pp. 799–808.

Edited by R. Amils & A. Ballester. Amsterdam: Elsevier.Clark, D. A. & Norris, P. R. (1996). Acidimicrobium ferrooxidans gen. nov., sp. nov. mixed culture ferrous iron oxidation with Sulfobacillus species. Microbiology 141, 785–790.

Hallberg, K. B. & Johnson, D. B. (2001). Biodiversity of acidophilic prokaryotes. Adv Appl Microbiology 49, 37–84.

Hallberg, K. B. & Lindstro¨ m, E. B. (1994). Characterization of Thiobacillus caldus sp. nov., a moderately thermophilic acidophile. Microbiology 140, 3451–3456.

Hallberg, K. B., Johnson, D. B. & Williams, P. A. (1999). A novel metabolic phenotype among acidophilic bacteria: aromatic degradation and the potential use of these microorganisms for the treatment of wastewater containing organic and inorganic pollutants. In Bio hydrometallurgy and the Environment. Toward the Mining of the 21st Century, Process Metallurgy 9A, pp. 719–728. Edited by R. Amils & A. Ballester. Amsterdam: Elsevier.

Hallberg, K. B., Coupland, K., Kimura, S. & Johnson, D. B. (2006). Macroscopic ''acid streamer'' growths in acidic, metal-rich mine waters in north Wales consist of novel and remarkably simple bacterial communities. Appl Environ Microbiol 72, 2022–2030.

Harvey, T. J. & Bath, M. (2007). The GeoBiotics GEOCOAT technology – progress and challenges. In Biomining, pp. 113–138. Edited by D. E. Rawlings & D. B. Johnson. Heidelberg: Springer.

Hawkes, R. B., Franzmann, P. D. & Plumb, J. J. (2006). Moderate thermophiles including ''Ferroplasma cyprexacervatum'' sp. nov., dominate an industrial scale chalcocite heap bioleaching operation. Hydrometallurgy 83, 229–236.

Johnson, D. B., Rolfe, S., Hallberg, K. B. & Iversen, E. (2001a). Isolation and phylogenetic characterisation of acidophilic microorganisms indigenous to acidic drainage waters at an abandoned Norwegian copper mine. Environ Microbiol 3, 630–637.

Johnson, D. B., Bacelar-Nicolau, P., Okibe, N., Yahya, A. & Hallberg, K. B. (2001b). Role of pure and mixed cultures of Gram-positive eubacteria in mineral leaching. In Biohydrometallurgy: Fundamentals, Technology and Sustainable

Development, Process Metallurgy 11A, pp. 461–470. Edited by V. S. T. Ciminelli & O. Garcia, Jr. Amsterdam: Elsevier.

Johnson, D. B., Okibe, N. & Roberto, F. F. (2003). Novel thermoacidophiles isolated from geothermal sites in Yellowstone National Park: physiological and phylogenetic characteristics. Arch Microbiology 180, 60–68.

Johnson, D. B., Okibe, N. & Hallberg, K. B. (2005). Differentiation and identification of iron-oxidizing acidophilic bacteria using cultivation techniques and amplified ribosomal DNA restriction enzyme analysis (ARDREA). J Microbiol Methods 60, 299–313.

Johnson, D. B., Stallwood, B., Kimura, S. & Hallberg, K. B. (2006). Characteristics of Acidicaldus organovorus, gen. nov., sp. nov.; a novel thermo-acidophilic heterotrophic proteobacterium. Arch Microbiology 185, 212–221.

Kinnunen, H.-M. & Puhakka, J. A. (2004). High-rate ferric sulfate generation by a Leptospirillum ferriphilum-dominated biofilm and the role of jarosite in biomass retainment in a fluidized-bed reactor. Biotechnol Bioeng 85, 697–705.

Logan, T. C., Seal, T. & Brierley, J. A. (2007). Whole-ore heap biooxidation of sulfidic gold-bearing ores. In Biomining, pp. 113–138. Edited by D. E. Rawlings & D. B. Johnson. Heidelberg: Springer.

Marsh, R. M. & Norris, P. R. (1983). The isolation of some thermophilic, autotrophic, iron- and sulphur-oxidizing bacteria. FEMS Microbiol Lett 17, 311–315.

Mikkelsen, D., Kappler, U., McEwan, A. G. & Sly, L. I. (2006). Archaeal diversity in two thermophilic chalcopyrite bioleaching reactors. Environ Microbiol 8, 2050–2055.

Norris, P. R., Clark, D. A., Owen, J. P. & Waterhouse, S. (1996). Characteristics of Sulfobacillus acidophilus sp. nov. and other moderately thermophilic mineral-sulphide-oxidizing bacteria. Microbiology 141, 775–783.

Norris, P. R., Burton, N. P. & Foulis, N. A. M. (2000). Acidophiles in bioreactor mineral processing. Extremophiles 4, 71–76. Okibe, N. & Johnson, D. B. (2004).

Okibe, N., Gericke, M., Hallberg, K. B. & Johnson, D. B. (2003). Enumeration and characterization of acidophilic microorganisms isolated from a pilot plant stirred tank bioleaching operation. Appl Environ Microbiol 69, 1936–1943.

Tuffin, I. M., Hector, S. B., Deane, S. M. & Rawlings, D. E. (2006). The resistance determinants of a highly arsenic resistant strain of Leptospirillum ferriphilum isolated from a commercial biooxidation tank. Appl Environ Microbiol 72, 2247–2253.

Waksman, S. A. & Joffe, J. S. (1921). Acid production by a new sulfur-oxidizing bacterium. Science 53, 216.

Yahya, A. & Johnson, D. B. (2002). Bioleaching of pyrite at low pH and low redox potentials by novel mesophilic Gram-positive bacteria. Hydrometallurgy 63, 181–188.

11

Biodegradable Plastics

Our whole world seems to be wrapped in plastic. Almost every product we buy most of the food we eat and many of the liquids we drink come encased in plastic. Packaging is the largest market for plastics, accounting for over a third of the consumption of raw plastic materials. Plastics are used on a daily basis throughout the world. The word plastic is a common term that is used for many materials of a synthetic or semi-synthetic nature. The term was derived from the Greek plastikos, which means "fit for molding." Plastics are a wide variety of combinations of properties when viewed as a whole. They are used for shellac, cellulose, rubber, and asphalt. We also synthetically manufacture items such as clothing, packaging, automobiles, electronics, aircrafts, medical supplies, and recreational items. The list could go on and on and it is obvious that much of what we have today would not be possible without plastics.

One way plastics changed the world was in cost. It was so much cheaper to manufacture than other materials and the various ways it could be used was staggering. The properties of plastic can also be altered. This can occur by modifying the polymers from the original, or changed by additives, colorants, reinforcement, or fillers. Plastics have become a staple product in the world; the only real concern is how they will affect the environment. Latest news speaks of plastics that release toxins; there are litter concerns and the effect on landfills and waterways. While costs had been low, there has been a steady rise due to the cost of the fuels needed for its production. The solution is still unknown, but science is working on the problem and has found some promising alternatives such as fructose and oil shale. Only time will tell what the future of plastics will be, but it is unlikely that the world will ever be without them.

ENVIRONMENTAL IMPACT OF PLASTICS

From cell phones and computers to bicycle helmets and hospital IV bags, plastic has molded society in many ways that make life both easier and safer. But the synthetic material also has left harmful imprints on the environment and perhaps human health.

Substantial quantities of plastic have accumulated in the natural environment and in landfills. Around 10 per cent by weight of the municipal waste stream is plastic. Discarded plastic also contaminates a wide range of natural terrestrial, freshwater and marine habitats, with newspaper accounts of plastic debris on even some of the highest mountains. There are accounts of inadvertent contamination of soils with small plastic fragments as a consequence of spreading sewage sludge (Zubris & Richards 2005), of fragments of plastic and glass contaminating compost prepared from municipal solid waste (Brinton 2005) and of plastic being carried into streams, rivers and ultimately the sea with rain water and flood events.

Evidence is mounting that the chemical building blocks that make plastics so versatile are the same components that might harm people and the environment. And its production and disposal contribute to an array of environmental problems, too. For example:

1) Chemicals added to plastics are absorbed by human bodies. Some of these compounds have been found to alter hormones or have other potential human health effects.
2) Plastic debris, laced with chemicals and often ingested by marine animals, can injure or poison wildlife.
3) Floating plastic waste, which can survive for thousands of years in water, serves as mini transportation devices for invasive species, disrupting habitats.
4) Plastic buried deep in landfills can leach harmful chemicals that spread into groundwater.
5) Around 4 percent of world oil production is used as a feedstock to make plastics, and a similar amount is consumed as energy in the process.
6) People are exposed to chemicals from plastic multiple times per day through the air, dust, water, food and use of consumer products.
7) Phthalates are used as plasticizers in the manufacture of vinyl flooring and wall coverings, food packaging and medical devices. Eight out of every ten babies, and nearly all adults, have measurable levels of phthalates in their bodies.
8) Bisphenol A (BPA), found in polycarbonate bottles and the linings of food and beverage cans, can leach into food and drinks. The U.S. Centers for Disease Control and Prevention reported that 93 percent of people had detectable levels of BPA in their urine.
9) Plastic also serves as a floating transportation device that allows alien species to hitchhike to unfamiliar parts of the world, threatening biodiversity. Global warming further aids the process by making

previously inhospitable areas like the Arctic livable for invasive species, which can be detrimental to local species.

10) Production of plastics is a major user of fossil fuels. Eight percent of world oil production goes to manufacturing plastics.

Global Issue

Marine litter and in particular plastic waste, is a global problem. The vast majority of plastic waste is destined for a landfill site which limits the impact through 'containment' however does not solve the problem. A significant proportion of plastic gets into the water course and eventually ends up in the oceans. As might be expected the plastic waste on the coastlines is more prevalent around more populated coastal areas. However, once the plastic waste enters the oceans it is influenced by global currents that distribute it around the world.

Plastic has been found in all of the major oceans, not just areas of human habitation, often travelling vast distances. It does not respect international boundaries and has invaded even the most remote places. 46% of plastics float (EPA 2006) and it can drift for years before eventually concentrating in the ocean gyres. These ocean currents create zones of convergence where large amounts of plastic waste accumulates; much of this is particulate plastic that has been broken down through wind, wave and UV action over a period of time.

Impact on Marine Ecosystems

Plastic Pollution is having a significant environmental impact particularly on marine life and coastlines. There are three major impacts on marine ecosystems:

Entanglement

Over 250 species have been known to have ingested or become entangled in plastic (Laist, 1997). Entanglement rates of up to 7.9% have been discovered in some species of seals and sea lions (Allsopp et al). A UNEP report estimates that around 130,000 cetaceans are caught in nets each year (US EPA, 1992).

Ingestion

Over 100 species of sea birds are known to ingest plastic artifacts (Laist, 1997). Around 95% of Fulmers have plastic in their stomachs that affect them in chemical and mechanical ways. 31 species of marine mammals are known to have ingested marine plastic (Allsopp et al).

Transport of Invasive Species

The increase in marine litter, in particular plastics has resulted in a corresponding increase in species invasion (Allsopp et al). Man-made litter has resulted in a significant increase in the opportunities for the transportation of alien species. 'Biotic mixing' as a result of human activities is becoming a widespread problem (Barnes, 2002). The hard surfaces of plastic debris are providing an attractive and alternative substrate for a number of organisms. The introduction of non-endemic species can have a catastrophic impact on indigenous species and biodiversity and the increase in synthetic and non-biodegradable material pollution will accelerate the process (Gregory, 2009).

The fact is that it doesn't matter where you live; plastic waste is pervasive, pernicious and persistent. It reaches every part of the planet and it is all of our responsibilities to resolve this issue.

Impact on Human Health

Plastic waste has serious impact on the human health. It results in increase in various deadly ailments like:

1) Cancer
2) Diabetes
3) Low sperm count
4) Altered immune systems
5) Genital defects
6) Endocrine disruptors
7) Rheumatoid arthritis
8) Endometriosis
9) Low birth weights
10) Developmental problems in children like lowered IQ, Lowered reading ages, affected social skills, behavioral problems memory and attention problems etc.

BIODEGRADABLE PLASTICS

In an effort to overcome these shortcomings, biochemical researchers and engineers have long been seeking to develop biodegradable plastics that are made from renewable resources, such as plants. The term biodegradable means that a substance is able to be broken down into simpler substances by the activities of living organisms, and therefore is unlikely to persist in the environment **(Fig 11.1)**. There are many different standards used to measure biodegradability, with each country having its own. The requirements range from 90 per cent to 60 per cent decomposition of the product within 60 to 180 days of being placed in a standard composting environment.

The reason traditional plastics are not biodegradable is because their long polymer molecules are too large and too tightly bonded together to be broken apart and assimilated by decomposer organisms. However, plastics based on natural plant polymers derived from wheat or corn starch have molecules that are readily attacked and broken down by microbes.

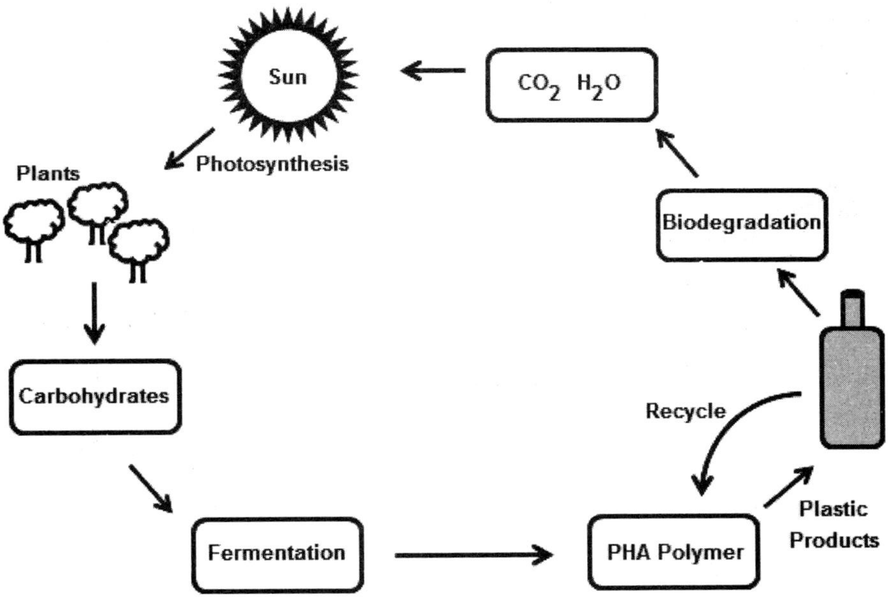

Fig. 11.1: Carbon Cycle of Bioplastics

Biodegradable plastics are made from all natural plant materials. These can include: corn oil, orange peels, starch, and plants. Traditional plastic is made with chemical fillers that can be harmful to the environment when released when the plastic is melted down. With biodegradable plastic, you get a substance made from natural sources that does contain these chemical fillers; they do not pose the same risk to the environment. The process of making biodegradable plastics begins with the melting down of all the materials. That mixture is then poured into molds of various shapes such as plastic water bottles and utensils. Biodegradable plastics are generally plant-derived polymers, produced by converting plant sugars into plastic, producing the polymer inside a microorganism (bacteria), or growing the plastic inside the leaves or stalk of corn or other crops.

Plastics can be Produced from Starch (PLA)

Starch is a natural polymer. It is a white, granular carbohydrate produced by plants during photosynthesis and it serves as the plant's energy store. Cereal plants and tubers normally contain starch in large proportions. Starch can be processed directly into a bioplastic but, because it is soluble in water, articles made from starch will swell and deform when exposed to moisture, limiting its use. This problem can be overcome by modifying the starch into a different polymer. First, starch is harvested from corn, wheat or potatoes, and then microorganisms transform it into lactic acid, a monomer. Finally, the lactic acid is chemically treated to cause the molecules of lactic acid to link up into long chains or polymers, which bond together to form a plastic called polylactide (PLA)(**Fig 11.2**).

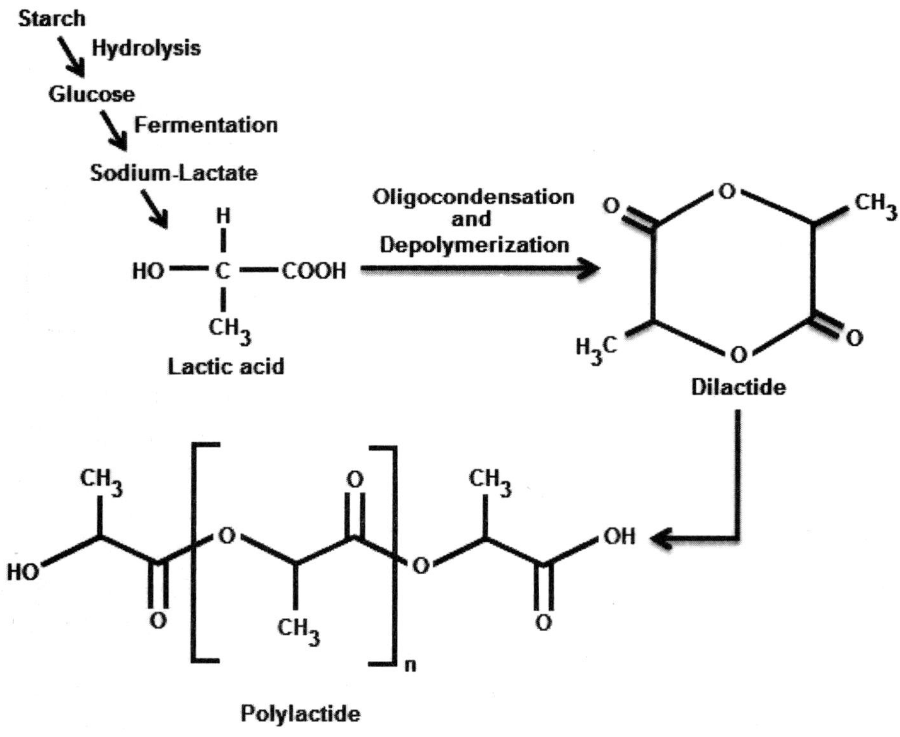

Fig. 11.2: Plastic (PLA) Production from Starch

PLA can be used for products such as plant pots and disposable nappies. It has been commercially available since 1990, and certain blends have proved successful in medical implants, sutures and drug delivery systems because of their capacity to dissolve away over time. However, because PLA is significantly more expensive than conventional plastics it has failed to win widespread consumer acceptance.

Plastics from Cellulose

Cellulose is an important structural component of the primary cell wall of green plants, many forms of algae and the oomycetes. Some species of bacteria secrete it to form biofilms. Cellulose is the most abundant organic polymer on Earth. The cellulose content of cotton fiber is 90%, that of wood is 40–50% and that of dried hemp is approximately 45%. Cellulose is an organic compound with the formula $(C_6H_{10}O_5)_n$, a polysaccharide consisting of a linear chain of several hundred to over ten thousand $\beta(1\rightarrow4)$ linked D-glucose units. Cellulose is mainly used to produce paperboard and paper. Smaller quantities are converted into a wide variety of derivative products such as cellophane and rayon. Conversion of cellulose from energy crops into biofuels such as cellulosic ethanol is under investigation as an alternative fuel source. Cellulose for industrial use is mainly obtained from wood pulp and cotton.

Cellulose Acetate

Cellulose acetate fiber is one of the earliest synthetic fibers and is based on cotton or tree pulp cellulose ("biopolymers"). These "cellulosic fibers" have been replaced in many applications by cheaper petro-based fibers (nylon and polyester) in recent decades. Acetate is a very valuable manufactured fiber that is low in cost and has good draping qualities. Properties of acetate have promoted it as the “beauty fiber”. Acetate is used in fabrics such as satins, brocades, and taffetas to accentuate luster, body, drape and beauty.

Acetate is derived from cellulose by deconstructing wood pulp into purified fluffy white cellulose. In order to get a good product special qualities of pulps - dissolving pulps - are used. A common problem with these is that the reactivity of the cellulose is uneven, and the quality of the cellulose acetate will sometimes be impacted. The cellulose is then reacted with acetic acid and acetic anhydride in the presence of sulfuric acid. It is then put through a controlled, partial hydrolysis to remove the sulfate and a sufficient number of acetate groups to give the product the desired properties. The anhydroglucose unit is the fundamental repeating structure of cellulose and has three hydroxyl groups which can react to form acetate esters. The most common form of cellulose acetate fiber has an acetate group on approximately two of every three hydroxyls. This cellulose diacetate is known as secondary acetate, or simply as "acetate". Major industrial acetate fiber uses are as follows:

a) Apparel: buttons, sunglasses, linings, blouses, dresses, wedding and party attire, home furnishings, draperies, upholstery and slip covers.

b) High absorbency products: diapers and surgical products.

c) The original Lego bricks were manufactured from cellulose acetate from 1949 to 1963.

d) Award Ribbon: Rosettes for equestrian events, dog/cat shows, corporate awards, advertising and identification products all use cellulose acetate ribbon.

e) KEM brand playing cards, used at the World Series of Poker and in many poker rooms at major casinos, are made of cellulose acetate. Italian playing card manufacturer Modiano also makes a line of playing cards made of "acetate," though it is unclear whether this is true cellulose acetate.

Plastics from Chitin

Chitin is a polysaccharide. A polysaccharide is a polymer - a giant molecule consisting of smaller molecules of sugar strung together. Chitin can be described as a biopolymer composed of N-acetyl-D-glucosamine; a chemical structure very close to cellulose except that the hydroxyl group in C (2) of cellulose is being replaced by an acetamido group in chitin. It is one of the most abundant natural polymers on Earth, found in fungal cell walls, shells of marine invertebrates, and the exoskeletons of insects. The bioplastic created from chitin can be used in the form of ropelike fibers or films.

Chitosan - A Derivative of Chitin

Chitin is modified to chitosan **(Fig 11.3)** by the elimination of the N-acetyl groups. Chitosan is a cationic polysaccharide which opens up a wide range of application opportunities in an acid environment. Chitosan can be biologically degraded, is non-toxic, inhibits the growth of bacteria and fungi and also enhances wound healing. Many researchers and companies use chitosan in the treatment of water and wastewater, in agriculture, in the paper and textile industries as well as for the production of bioplastics.

Chitosan is a biological product with cationic (positive electrical charge) properties. It is of great interest, all the more so because most polysaccharides of the same types are neutral or negatively charged. By controlling the molecular weight, the degree of deacetylation and purity, it is possible to produce a broad range of chitosans and derivatives that can be used for industrial, dietary, cosmetic and biomedical purposes. Chitosan offers a natural alternative to the use of chemical products that are sometimes harmful to humans and their environment. Chitosan triggers the defensive mechanisms in plants (acting much like a vaccine in humans), stimulates growth and induces certain enzymes (synthesis of phytoalexins, chitinases, pectinases, glucanases, and lignin). This new organic control approach offers promise as a bio control tool. In addition to the growth-stimulation properties and fungi, chitosans are used for:

a) Seed-coating.

b) Frost protection.

c) Bloom and fruit-setting stimulation.

d) Timed release of product into the soil (fertilizers, organic control agents, nutrients).

e) Protective coating for fruits and vegetables.

Biodegradable Plastics from Microorganisms

Another way of making biodegradable polymers involves getting bacteria to produce granules of a plastic called polyhydroxyalkanoate (PHA) inside their cells. Bacteria are simply grown in culture, and the plastic is then harvested. Going one step further, scientists have taken genes from this kind of bacteria and stitched them into corn plants, which then manufacture the plastic in their own cells. PHA refers to polymers that are composed of hydroxy carbonic acids, i.e. carbonic acids with one or several additional OH groups. Some organisms are able to combine hydroxy carbonic acids to polyesters due to their specific chemical structures. Over 150 different PHAs are known, and there are good reasons for such a large variety. Cells produce hydroxy carbonic acids from standard intermediary products of the energy metabolism and fatty acid synthesis. When key components in the food supply are lacking, organisms and cells turn to alternative metabolic pathways and form PHAs. Specific metabolic pathways are chosen on the basis of the carbon source provided to the cells, which leads to the creation of PHAs with different chemical structures.

Polyhydroxyalkanoate or PHAs are linear polyesters produced in nature by bacterial fermentation of sugar or lipids. These plastics are biodegradable and are used in the production of bioplastics. They can be either thermoplastic or elastomeric materials, with melting points ranging from 40 to 180 °C. The mechanical and biocompatibility of PHA can also be changed by blending, modifying the surface or combining PHA with other polymers, enzymes and inorganic materials, making it possible for a wider range of applications. PHA polymers are thermoplastic, can be processed on conventional processing equipment, and are, depending on their composition, ductile and more or less elastic. They differ in their properties according to their chemical composition (homo-or co-polyester, contained hydroxy fatty acids).

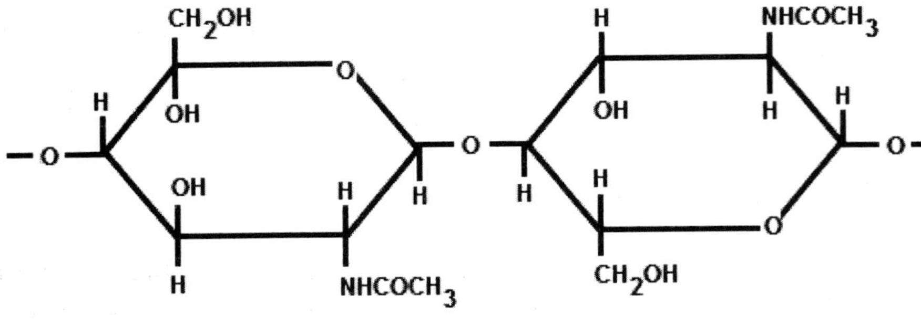

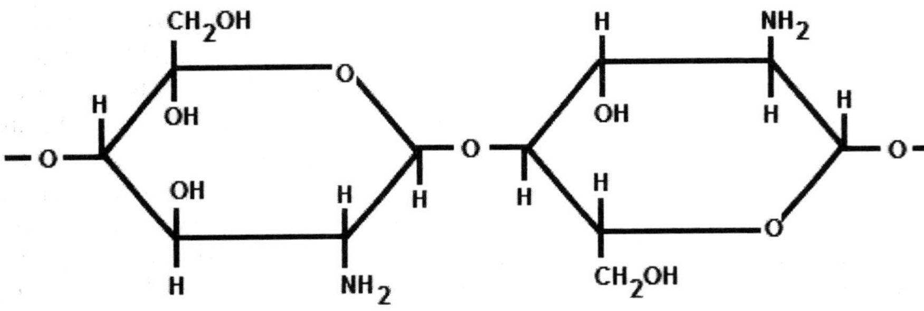

Fig. 11.3: Structure of Chitin and Chitosan

Polyhydroxybutyrate (PHB) is a polyhydroxyalkanoate (PHA), a polymer belonging to the polyesters class that are of interest as bio-derived and biodegradable plastics. The poly-3-hydroxybutyrate (P3HB) form of PHB is probably the most common type of polyhydroxyalkanoate, but other polymers of this class are produced by a variety of organisms: these include poly-4-hydroxybutyrate (P4HB), polyhydroxyvalerate (PHV), polyhydroxyhexanoate (PHH), polyhydroxyoctanoate (PHO) and their copolymers.

PHB is produced by microorganisms (such as *Ralstonia eutrophus* or *Bacillus megaterium*) apparently in response to conditions of physiological stress **(Fig 11.4)**. The polymer is primarily a product of carbon assimilation (from glucose or starch) and is employed by microorganisms as a form of energy storage molecule to be metabolized when other common energy sources are not available. Microbial biosynthesis of PHB starts with the condensation of two molecules of acetyl-CoA to give acetoacetyl-CoA which is subsequently reduced to hydroxybutyryl-CoA. This latter compound is then used as a monomer to polymerize PHB.

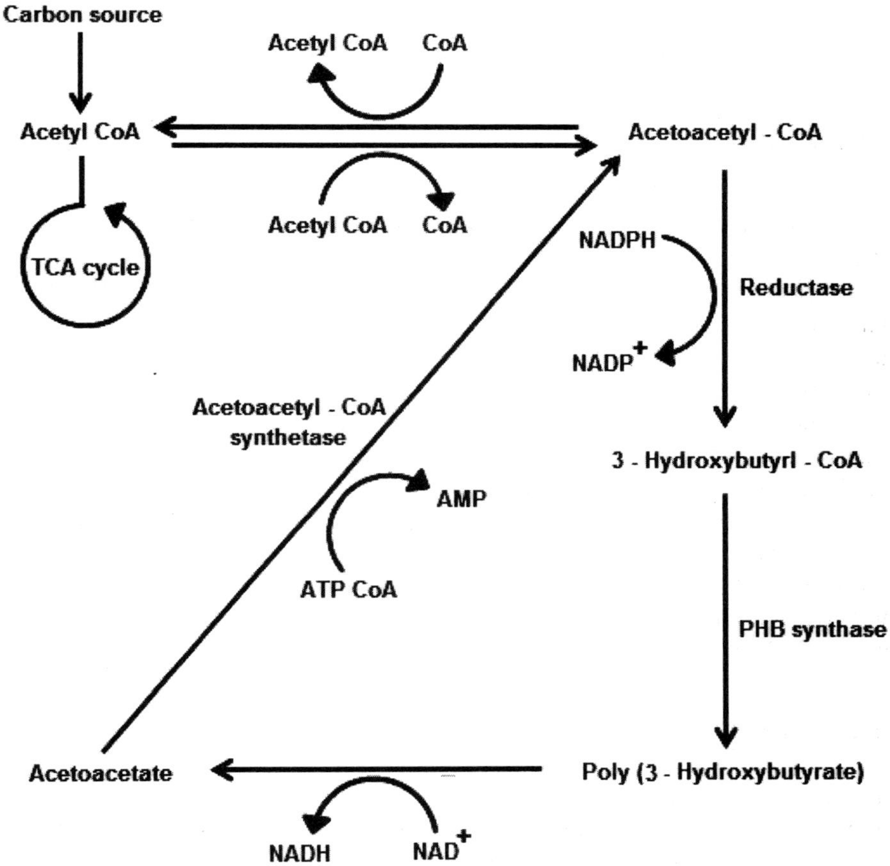

Fig. 11.4: Biosynthesis and Biodegradation of PHB

Properties of PHB

1) PHB is water insoluble and relatively resistant to hydrolytic degradation. This differentiates PHB from most other currently available biodegradable plastics, which are either water soluble or moisture sensitive.
2) Good oxygen permeability.
3) Good ultra-violet resistance but poor resistance to acids and bases.
4) Soluble in chloroform and other chlorinated hydrocarbons.
5) Biocompatible and hence is suitable for medical applications.
6) Melting point 175°C and glass transition temperature 2°C.
7) Tensile strength 40 MPa, close to that of polypropylene.

8) Sinks in water (while polypropylene floats), facilitating its anaerobic biodegradation in sediments.

9) Nontoxic.

10) Less 'sticky' when melted, making it a potentially good material for clothing in the future.

PHAs are biodegradable thermoplastics that are synthesized by many different types of bacteria. When bacteria develop in nutrient-deficient environments, bacteria create PHAs as food and energy reserves, which are then stored as insoluble granules in the cytoplasm. Depending on their molecular composition, PHAs have varying physical properties, but all PHAs biodegrade in carbon dioxide and water. PHAs can degrade in either aerobic or anaerobic environments through thermal degradation or enzymatic hydrolysis. Bacteria, algae, and fungi use extracellular enzymes to depolymerize, or break down, PHAs and, through a process known as mineralization, absorb the remaining fragments for use as minerals. Manufacturers can alter the properties of PHAs by changing their structure or composition. PHAs have a variety of applications in water-resistant surfaces, binders, synthetic paper, medical devices, electronic parts, food packaging, and agriculture.

REGULAR PLASTIC vs BIODEGRADABLE PLASTIC

After formation, regular plastics hold carbon. When they are disposed of and begin to decompose or when they are melted, that carbon is then released into the atmosphere. Biodegradable plastics do not release carbon, because no carbon is involved in the manufacturing process.

Methane and other forms of pollutants could also be released from traditional plastic when they are recycled and burned. This is not the case with biodegradable plastics, which do not contain those polluting materials.

One of the many positive aspects of biodegradable plastics is that they are able to be broken down by naturally occurring bacteria, which again will be beneficial to the environment.

Aside from a slightly higher cost to produce, biodegradable plastics hold many advantages over standard plastics, with a lesser impact on the environment being one of its greatest advantages.

Production Cost and Scope of Biodegradable Plastics

PHAs currently have economic drawbacks that limit their use. Until recently, PHAs have had high production costs, low yields, and limited availability.

Consequently, they have not been able to displace the petroleum-based plastics that manufacturers can cheaply create in bulk.

One method of reducing the cost of production is to create a polymer blend of PHAs with renewable materials, such as starch or cellulose, a technique that requires a smaller amount of PHA per unit. The blends still have properties that are easy to modify, providing a viable, less expensive alternative to pure PHAs. Blends also degrade better than PHAs without renewable materials. When PHAs are blended with hydrophilic polymers, more water can penetrate the plastic and increase the efficiency of degradation.

TRANSGENIC PLANTS PRODUCING PHA

Production costs are major impediments to the marketability of biodegradable plastics. Researchers have experimented with various methods of production, but one of the most promising techniques is cultivating PHAs in transgenic plants. Synthesizing PHAs through bacterial fermentation costs five times as much as the production of petroleum-based plastics because of low yields per bacterium. By using transgenic plants to produce PHAs, net yields increase at a lower cost. The plants can grow PHAs by redirecting cytoplasmic acetyl-CoA present in the plant to produce PHB. However, redirecting cytoplasmic acetyl-CoA stunts plant growth and negatively affects yield of both PHB and the plant itself. To avoid this side effect, researchers have instead targeted PHB production to specific areas of the plant with existing high levels of acetyl-CoA, such as chloroplasts. When synthesized in this manner, PHB yields make up 15% of the plant's dry weight, dramatically lowering the cost of production. Research is being done on various plants like *Arabidopsis thaliana*, Tobacco (*Nicotiana tabacum*), Rapeseed (*Brassica napus*), Cotton (*Gossypium hirsutum*), Alfalfa (*Medicago sativa*), Oil palms (*Elaeis guineensis* and *E. oleifera*), Flax (*Linum usitatissimum*) for the scope of production of PHB in them.

FUTURE OF BIODEGRADABLE PLASTICS

Researchers have worked on developing biodegradable plastics with the hopes of bettering the environment, but the production methods and applications of biodegradable plastics could still be detrimental to environmental and human health. There are certain issues to be addressed before replacement of petroleum based plastics with biodegradable plastics:

a) The waste management infrastructure currently recycles regular plastic waste, incinerates it, or places it in a landfill. Mixing biodegradable plastics into the regular waste infrastructure poses some dangers to the environment. Biodegradable plastics behave differently when recycled, and have the potential to negatively influence to human health.

b) To be effective in food packaging, plastics must exhibit gas permeability, chemical resistance, and tensile strength. If the food packaging materials are recycled, their physical properties could change, allowing degraded chemical compounds and external contaminants to enter the food. On top of that, plastics contaminated with food are difficult to recycle, and blended plastics sometimes leave behind starch residues that can further contaminate the recycling process.

c) Another option for biodegradable plastic waste is incineration with energy capture, so that the energy that goes into producing the plastic could be reclaimed during decomposition. However, incineration of biodegradable plastics does not create any more energy than petroleum-based plastics, so the environmental effects of the two are roughly equivalent.

d) The next option is land filling biodegradable plastics. However, when biodegradable plastics decompose, they produce methane gas, a major contributor to global warming. While methane gas could be collected and used as an energy source, capturing that energy would be another expense, and some of the gas would still escape.

Thus, the biodegradable nature of these plastics poses economic and ecological problems in the current waste management infrastructure. Biodegradable plastics offer a promising alternative to petroleum-based plastics. While petroleum-based products use oil in their manufacturing and take up space in landfills, biodegradable plastics can be synthesized in bacteria or plants and have the potential to be disposed of in a way that is less damaging to the environment. Biodegradable plastics have a variety of applications, from agriculture and food packaging to biomedical devices and tableware. The major obstacles to replacing petroleum-based plastics with biodegradable plastics are high costs and low yields associated with existing methods of biodegradable plastic production. With more research into plant-based manufacturing systems, these obstacles are being overcome. Finally, the last obstacle to surmount is the proper disposal of biodegradable plastics. In order for biodegradable plastics to be effectively disposed of, the current waste management infrastructure must change, or methods with less economic and environmental costs must be developed.

REFERENCES

Ayorinde, F.O., Saeed, K.A., Price, E., Morrow, A., Collins, W.E., McInnis, F., Pollack, S.K. and Eribo, B.E. (1998). Production of poly-(b-hydroxybutyrate) from saponified Vernonia galamensis oil by Alcaligenes eutrophus. Journal of Industrial Microbiology and Biotechnology, 21, 46 – 50.

Biodegradable Plastics - Developments and Environmental Impacts. Nolan-ITU Pty Ltd and ExcelPlas Australia. Report prepared for Australia Department of Environment and Water Resources. October 2002.

Biodegradable Products Institute 331 W. 57th St. Suite 415, New York, N.Y. USA

BPI assessment of oxo-degradable films. Biodegradable Products Institute. Personal communication, Steve Mojo. May, 2007.

Cerdan, C., Gazulla, C., Raugei, M., Martinez, E. & Fullana-iPalmer, P. (2009) Proposal for new quantitative eco-design indicators: a first case study. Journal of Cleaner Production 17:1638-1643

Cho, H.S., Moon, H.S., Kim, M., Nam, K. & Kim, J.Y. (2011) Biodegradability and biodegradation rate of poly(caprolactone)-starch blend and poly(butylene succinate) biodegradable polymer under aerobic and anaerobic environment. Waste Management 31:475-480.

Chi, Z., Liu, G., Wang, F., Ju, L., Zhang, T. (2009). Saccharomycopsis fibuligera and its applications in biotechnology. Biotechnology Advances, 1, 1-9.

Confused by the term biodegradable and biobased. Biodegradable Products Institute. EPI-Global. http://www.epi-global.com/en/Index-e.htm. 2005.

Ekborg, N.A., Gonzalez, J.M., Howard, M.B., Taylor, L.E., Hutcheson, S.W. and Weiner, R.M. (2005). Saccharophagus degradans gen. nov., sp. nov., a versatile marine degrader of complex polysaccharides. International Journal of Systematic and Evolutionary Microbiology, 55, 1545-1549.

European Bioplastics (2008) Position Paper: Life Cycle Assessment of Bioplastics. Downloadable from http://www.european-bioplastics.org/index.php?id=191

European Bioplastics (2009) Position Paper: "Oxo-biodegradable" plastics. Downloadable from http://www.europeanbioplastics.org/index.php?id=191

European Plastics Recyclers Association (2010) How to increase the mechanical recycling of post-consumer plastics. Strategy paper of the European Plastics Recyclers Association. Downloadable from http://www.plasticsrecyclers.eu/.../

HowIncreaseRecycling/1265184667EUPR How To Increase Plastics Recycling FINAL low.pdf

Hendrickson, C.T., Matthews, D.H., Ashe, M., Jaramillo, P. & McMichael, F.C. (2010) Reducing environmental burdens of solidstate lighting through end-of-life design. Environmental Research Letters 5. Doi: 10.1088/1748-9326/5/1/014016

Hopewell, J., Dvorak, R. & Kosior, E. (2009). Plastics recycling: challenges and opportunities. Philosophical Transactions of the Royal Society B 364: 2115-2126

Kocer, H., Borcakli, M. and Demirel, S. (2003). Production of bacterial polyesters from some various new substrates by Alcaligenes eutrophus and Pseudomonas oleovorans. Turk J Chem, 27, 365 – 373.

Lazarevic, D., Aoustin, E., Buclet, N. & Brandt, N. (2010) Plastic waste management in the context of a European recycling society: Comparing results and uncertainties in a life cycle perspective. Resources, Conservation and Recycling 55:246-259

Lemmel, S.A., Heimsch, R.C. and Korus, R.A. (1980). Kinetics of growth and amylase production of Saccharomycopsis fibuligera on potato processing wastewater. Applied and Environmental Microbiology, 39, 1-7.

Loo, C.-Y. and Sudesh, K. (2007). Polyhydroxyalkanoates: Bio-based microbial plastics and their properties. Malaysian Polymer Journal, 2(2), 31-57.

Madison, L.L and Huisman, G.W. (1999). Metabolic engineering of poly(3-hydroxyalkanoates): from DNA to plastic. Microbiology and Molecular Biology Reviews, 63, 21–53.

Sudesh, K., Abe, H. and Doi, Y. (2000). Synthesis, structure and properties of polyhydroxyalkanoates: biological polyesters. Progress in Polymer Science, 25, 1503-1555.

Suvorov, M., Kumar, R., Zhang, H. and Hutcheson, S. (2011). Novelties of the cellulolytic system of a marine bacterium applicable to cellulosic sugar production. Biofuels, 2(1),1-12.

Tabone, M.D., Cregg, J.J., Beckman, E.J. & Landis, A.E. (2010) Sustainability Metrics: Life Cycle Assessment and Green Design in Polymers. Environmental Science & Technology. 44 (21): 8264–8269

Watson, B.J., Zhang, H., Longmire, A.G., Moon, Y.H. and Hutcheson, S.W. (2009). Processive endoglucanases mediate degradation of cellulose by Saccharophagus degradans. Journal of Bacteriology, 191(18), 5697-5705.

Yamashita, I., Itoh, T. and Fukui, S. (1985a). Cloning and expression of the Saccharomycopsis fibuligera α-amylase gene in Saccharomyces cerevisiae. Agricultural and Biological Chemistry, 49, 3089-3091.

Yamashita, I., Itoh, T. and Fukui, S. (1985b). Cloning and expression of the Saccharomycopsis fibuligera α-amylase gene in Saccharomyces cerevisiae. Applied Microbiology and Biotechnology, 23, 130-133.

12

Biofertilizers

Plants require optimal amounts of available nutrients for normal growth. These nutrients can come from several sources, including soil organic matter, native soil minerals, organic materials that are added to the soil (e.g., animal manures), air (e.g., legumes), and commercial fertilizers. When a soil is not capable of supplying enough nutrients to meet crop/plant requirements, commercial fertilizers can be added to supply the needed nutrients. There are numerous types of fertilizers that can be used to supply primary, secondary, or micronutrients.

A fertilizer is a material that furnishes one or more of the chemical elements necessary for the proper development and growth of plants. The most important ones are chemical or mineral fertilizers, manures, and plant residues. A chemical fertilizer is a material produced by industrial processes with the specific purpose of being used as a nutrient for plants. Fertilizers are essential in today's agricultural system to replace the elements extracted from the soil in the form of food and other agricultural products.

Most fertilizers that are commonly used in agriculture contain the three basic plant nutrients: nitrogen, phosphorus, and potassium. Some fertilizers also contain certain "micronutrients," such as zinc and other metals that are necessary for plant growth. Materials that are applied to the land primarily to enhance soil characteristics (rather than as plant food) are commonly referred to as soil amendments.

Fertilizer can be made by:

1) Natural processes, such as composting.
2) Chemical processes in manufacturing.
3) Mining and processing of minerals.

In order to grow, thrive, and multiply, plants need a variety of nutrients:

a) Common Elements: Carbon, hydrogen, and oxygen: easily obtained from air and water.

b) Primary Nutrients: Nitrogen, phosphorus, and potassium: sometimes referred to by their chemical symbols N-P-K. They are considered the "primary nutrients" and are usually the main active ingredients in fertilizer, indicated in order by the three numbers on the package label 21-7-14. These numbers refer to the percentage of nitrogen, phosphorus, and potassium in the package.

c) Secondary Nutrients: Calcium, sulfur, and magnesium along with trace elements such as boron, zinc, copper, chlorine, molybdenum, manganese, and iron, are also important to plants and may be present in fertilizer

TYPES OF FERTILIZERS

Straight Fertilizers

Straight fertilizers are those that supply only one of the major nutrients: nitrogen, phosphorus, potassium or magnesium. The amount of nutrient in the fertilizer is expressed as a percentage.

Compound Fertilizers

Compound fertilizers are those that supply two or more of the nutrients nitrogen, phosphorus and potassium. The nutrient content expressed as for straight fertilizers is, by convention, written on the bag in the order nitrogen, phosphorus and potassium.

Fertilizer regulations require that further details of trace elements, pesticide content and phosphorus solubility should appear where applicable on the invoice. Fertilizers and manures are available in many different forms. Generally the term organic implies that the fertilizer is derived from living organisms, whereas inorganic fertilizers are those derived from non-living material.

Chemical Fertilizers

Chemical fertilizers (also called inorganic, synthetic, artificial, or manufactured) have been refined to extract nutrients and bind them in specific ratios with other chemical fillers. These products may be made from petroleum products, rocks, or even organic sources. Some of the chemicals may be naturally occurring, but the difference is that the nutrients in chemical fertilizers are refined to their pure state and stripped of substances that control their availability and breakdown, which rarely occurs in nature.

Advantages of Chemical Fertilizer

a) Since nutrients are available to the plants immediately, improvement occurs in days.

b) They are highly analyzed to produce the exact ratio of nutrients desired.

c) Standardized labeling makes ratios and chemical sources easy to understand.

d) They're inexpensive.

Disadvantages of Chemical Fertilizer

a) Chemical fertilizers are primarily made from nonrenewable sources, including fossil fuels.

b) They grow plants but do nothing to sustain the soil. The fillers do not promote life or soil health, and even packages labeled "complete" do not include the decaying matter necessary to improve soil structure. In fact, chemical fertilizers don't replace many trace elements that are gradually depleted by repeated crop plantings, resulting in long-term damage to the soil.

c) Because the nutrients are readily available, there is a danger of over fertilization. This not only can kill plants but upset the entire ecosystem.

d) Chemical fertilizers tend to leach, or filter away from the plants, requiring additional applications.

e) Repeated applications may result in a toxic buildup of chemicals such as arsenic, cadmium, and uranium in the soil. These toxic chemicals can eventually make their way into fruits and vegetables.

f) Long-term use of chemical fertilizer can change the soil pH, upset beneficial microbial ecosystems, increase pests, and even contribute to the release of greenhouse gases.

Organic Fertilizer

The words "organic" or "natural" simply means that the product is only minimally processed, and the nutrients remain bound up in their natural forms, rather than being extracted and refined. Organic fertilizer is usually made from plant or animal waste or powdered minerals. Examples include manure and compost, as well as bone and cottonseed meal. They are usually sold as "soil conditioners" rather than as fertilizer, because the nutrient ratios are difficult to guarantee. Organic fertilizers may be processed in a factory, or, in the case of manure and compost, at a farm.

EFFECTS OF FERTILIZERS ON ENVIRONMENT

Groundwater Pollution

Nitrate leaching through the soil can present a serious health hazard and contributes to soil acidification. When high rates of nitrogen are used or where clover grass pastures fix substantial nitrogen, especially on sandy or permeable soils, inevitably some nitrate is leached and may enter groundwater if there is a water table. If this groundwater is used for domestic supplies, the leaching presents a serious health hazard. Groundwater contamination has been linked to gastric cancer, goiter, birth malformations, and hypertension, testicular cancer and stomach cancer. Perhaps one of the scariest effects of chemical fertilizers is something called methemoglobinemia. In infants it is alternatively known as Blue Baby Syndrome. The risk most often occurs when infants are given formula reconstituted with nitrate contaminated water. The condition causes a decrease in oxygen in the blood and results in a blue-grey skin color, causes lethargy and/or irritability and can lead to coma or death.

Eutrophication

Eutrophication is the enrichment of water by the addition of nutrients. The extra nutrients encourage the growth of algal blooms, particularly in stagnant water. Blue–green algae may produce toxins poisonous to animals, including humans. For these algae to grow, phosphorus must be present in the water above a certain level.

Phosphorus may be introduced into waterways in run-off from pasture, forests and fertilized land and in drainage from irrigated land and urban areas. These sources, representing most of the total run-off, normally contribute low concentrations of phosphorus and are referred to as diffuse or non-point sources. Point sources, such as sewage effluent and drainage from dairies and feedlots, contribute smaller flows but contain much higher concentrations of phosphorus. These are frequently found to be the sources for most of the phosphorus found in waterways.

Soil Acidity

There are three major acidifying processes agricultural systems:

(i) Addition of nitrogen to the soil by fertilizer or fixation of atmospheric nitrogen, followed by loss of nitrate from the soil due to leaching or run-off.

(ii) Production of organic acids from decomposing organic matter.

(iii) Removal of alkaline products such as hay from the soil.

Heavy Metal Accumulation

Cadmium is present in tiny amounts (less than 0.5 mg/L) in the soil, and in larger amounts in rock phosphate. Plant uptake of cadmium is small, but when plants containing cadmium are grazed by livestock, the cadmium accumulates in and may reach very high concentrations. Uranium is another example of a contaminant often found in phosphate fertilizers. Eventually these heavy metals can build up to unacceptable levels and build up in vegetable produce.

Atmosphere

Methane emissions from crop fields are increased by the application of ammonium-based fertilizers; these emissions contribute greatly to global climate change as methane is a potent greenhouse gas. The use of fertilizers on a global scale emits significant quantities of greenhouse gas into the atmosphere. Emissions come about through the use of:

(i) Animal manures and urea, which release methane, nitrous oxide, ammonia, and carbon dioxide in varying quantities depending on their form (solid or liquid) and management (collection, storage, spreading).

(ii) Fertilizers that use nitric acid or ammonium bicarbonate, the production and application of which results in emissions of nitrogen oxides, nitrous oxide, ammonia and carbon dioxide into the atmosphere.

BIOFERTILIZERS

Sustainable crop production depends much on good soil health. Soil health maintenance warrants optimum combination of organic and inorganic components of the soil. Repeated use of chemical fertilizers destroys soil biota. In nature, there are a number of useful soil micro-organisms which can help plants to absorb nutrients. Their utility can be enhanced with human intervention by selecting efficient organisms, culturing them and adding them to soils directly or through seeds. The cultured micro organisms packed in some carrier material for easy application in the field are called bio-fertilizers.

Bio-fertilizers are living microorganisms of bacterial, fungal and algal origin. Their mode of action differs and can be applied alone or in combination. By systematic research, efficient strains are identified to suit to given soil and climatic conditions. Such strains have to be mass multiplied in laboratory and distributed to farmers. They are packed in carrier materials like peat, lignite powder in such a way that they will have sufficient shelf life.

Biofertilizer is a substance which contains living microorganisms which, when applied to seed, plant surfaces, or soil, colonizes the rhizosphere or the interior of the plant and promotes growth by increasing the supply or availability of primary nutrients to the host plant. Biofertilizers add nutrients through the natural processes of Nitrogen fixation, solubilizing phosphorus, and stimulating plant growth through the synthesis of growth promoting substances. Biofertilizers can be expected to reduce the use of chemical fertilizer and pesticides. The microorganisms in biofertilizers restore the soil's natural nutrient cycle and build soil organic matter. Through the use of biofertilizers, healthy plants can be grown while enhancing the sustainability and the health of soil.

The main sources of biofertilizers are bacteria, fungi, and cyanobacteria (blue-green algae). The most striking relationship that these have with plants is symbiosis, in which the partners derive benefits from each other. Very often microorganisms are not as efficient in natural surroundings as one would expect them to be and therefore artificially multiplied cultures of efficient selected microorganisms play a vital role in accelerating the microbial processes in soil.

Biofertilizers are Eco-friendly organic agro-input and more cost effective than chemical fertilizers. Biofertilizers like Rhizobium, Azotobacter, Azospirillum and blue green algae (BGA) are in use since long time ago. Rhizobium inoculants are used for leguminous crops. Azotobacter can be used with crops like wheat, maize, mustard, cotton, potato and other vegetable crops. Azospirillum inoculants are recommended mainly for sorghum, millets, maize, sugarcane and wheat. Blue green algae belonging to genera Nostoc, Anabaena, Tolypothrix and Aulosira fix atmospheric nitrogen and are used as inoculants for paddy crop grown both under upland and low land conditions.

Types of Biofertilizers

Use of biofertilizers is one of the important components of integrated nutrient management, as they are cost effective and renewable source of plant nutrients to supplement the chemical fertilizers for sustainable agriculture. Several microorganisms and their association with crop plants are being exploited in the production of biofertilizers. They can be grouped in different ways **(Table 12.1)** based on their nature and function.

1) Nitrogen fixing biofertilizers
2) Phosphate solubilizing biofertilizers
3) Phosphate mobilizing biofertilizers

4) Plant growth promoting biofertilizers
5) Biofertilizers for micro nutrients

Table 12.1: Different types of Biofertilizers

S.NO.	GROUPS	EXAMPLES
		Nitrogen Fixing Biofertilizers
1.	Free-living	*Azotobacter, Beijerinkia, Clostridium, Klebsiella, Anabaena, Nostoc,*
2.	Symbiotic	*Rhizobium, Frankia, Anabaena azollae*
3.	Associative Symbiotic	*Azospirillum*
		P Solubilizing Biofertilizers
1.	Bacteria	*Bacillus megaterium* var. *phosphaticum, Bacillus subtilis Bacillus circulans, Pseudomonas striata*
2.	Fungi	*Penicillium sp, Aspergillus awamori*
		P Mobilizing Biofertilizers
1.	Arbuscular mycorrhiza	*Glomus sp.,Gigaspora sp.,Acaulospora sp., Scutellospora sp. & Sclerocystis sp.*
2.	Ectomycorrhiza	*Laccaria sp., Pisolithus sp., Boletus sp., Amanita sp.*
3.	Ericoid mycorrhizae	*Pezizella ericae*
4.	Orchid mycorrhiza	*Rhizoctonia solani*
		Biofertilizers for Micro nutrients
1.	Silicate and Zinc solubilizers	*Bacillus* sp.
		Plant Growth Promoting Rhizobacteria
1.	Pseudomonas	*Pseudomonas fluorescens*

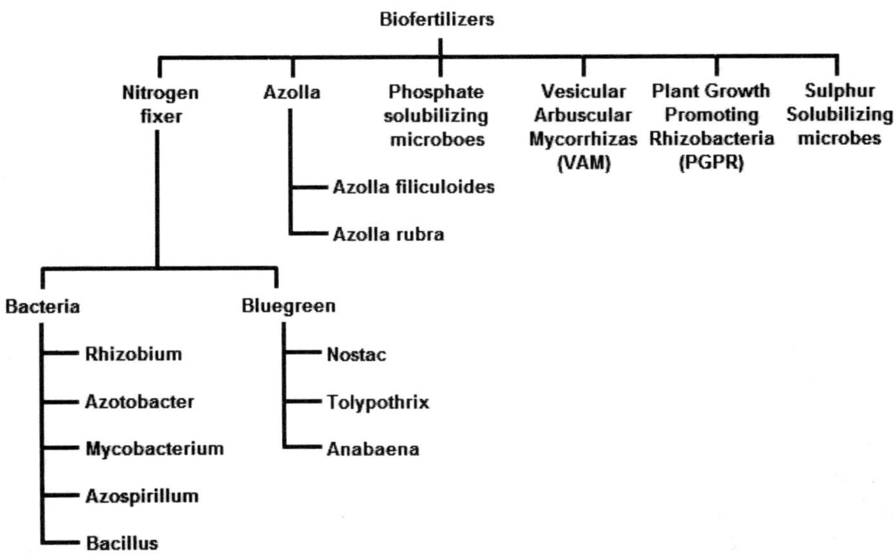

Fig. 12.1: Types of Biofertilizers

Bacteria as Biofertilizers

Many free-living and symbiotic bacteria fix atmospheric Nitrogen. Therefore, certain measures are adopted to increases number of such bacteria in soil which may increase the gross yield of nitrogen. The two methods, bacterization and green-manuring are the most widely used techniques.

Bacterization is a technique of seed-dressing with bacteria (as water suspension) for example, Azotobacter, Bacillus, Rhizobium etc. It has been proved that bacteria can successfully be established in root region of plants which in turn improve the growth of hosts. Bacterial fertilizers named 'azotobacterin' (containing cells of Azotobacter chroococcum) and ‘phosphobacterin' (containing cells of Baccilus megatehum var. phosphaticum) have been used in erstwhile U.S.S.R. and East European countries, respectively. These increased the crop yield about 10-20 per cent. Subsequently bacterization of seeds in Russia, Czechoslovakia, Rumania, Poland, Bulgaria, Hungary, England and India has clearly demonstrated the increase in crop yield such as wheat, barley, maize, sugarbeet, carrot, cabbage and potato. In rhizosphere, bacteria secrete growth substances and antibiotic secondary metabolites which contribute to seed germination and plant growth.

Green manuring is "a farming practice where a leguminous plant which has derived enough benefits from its association with appropriate species of Rhizobium is ploughed into the soil and then a non legume is grown and allowed to take the benefits of the already fixed nitrogen".

Some of these bacteria are discussed below:

Rhizobium

Rhizobium is a soil habitat bacterium, which can able to colonize the legume roots and fixes the atmospheric nitrogen symbiotically. The morphology and physiology of rhizobium will vary from free-living condition to the bacteroid of nodules. They are the most efficient bio fertilizer as per the quantity of nitrogen fixed concerned. They have seven genera and highly specific to form nodule in legumes, referred as cross inoculation group. The bacterium also produces enzymes (nitrogenase) that supply a constant source of reduced nitrogen to the host plant. It is suitable for Ground nut, Black gram, Green gram, Red gram, Cow pea, Bengal gram, Mustard, Soy bean, French bean, Cluster bean, Leguminous trees etc.

Azotobacter

Azotobacter is a genus of usually motile, oval or spherical bacteria that form thick-walled cysts and may produce large quantities of capsular slime. They are aerobic, free-living soil microbes which play an important role in the nitrogen cycle in nature, binding atmospheric nitrogen, which is inaccessible to plants, and releasing it in the form of ammonium ions into the soil. Apart from being a model organism, it is used by humans for the production of biofertilizers, food additives and some biopolymers. Azotobacter are Gram-negative bacteria. They are found in neutral and alkaline soils, in water and in association with some plants. Owing to its ability to fix molecular nitrogen and therefore increase the soil fertility and stimulate plant growth, Azotobacter are widely used in agriculture, particularly in nitrogen biofertilizers such as azotobacterin. Azotobacter chroococcum known to fix considerable quantity of nitrogen in the range of 20-40 kg of nitrogen / ha in the rhizosphere in non-leguminous crops. The lack of organic matter in the soil is a limited factor in the proliferation of Azotobacter in the soil. The bacterium induces hormones that help plants in better germination, early emergence and better root development.

Benefits:

1) Fixes atmospheric nitrogen which is a renewable source of energy.
2) Nitrogen fixed is immediately available to the plants.
3) Stimulates growth and induces green color characteristics of a healthy plant.
4) Improves soil properties and sustains soil fertility aids utilization of Potassium, Phosphorus and other nutrients.
5) Encourages plumpness and succulence of fruits and grains and increases protein percentage.
6) Increase the yield from 10% to 30%.
7) Improves the quality of product and thus fetches good price.

Azospirillum

Azospirillum lipofereum is a very useful soil and root bacterium. It is an associative symbiotic nitrogen fixing bacteria. It is found in the soil around plant roots and root surface. It also produces growth-promoting substances like indole acetic acid (IAA), gibberellins, pantothenic acid, thiamine and niacin and it promotes root proliferation and it improve the plant growth yield. It increases the rootlet density and root branching resulting in the increased uptake of mineral and water. Five species of azospirillum described to date are Brasilense, Lipoferum, Amazonense, Halopraeferens and Irakense. The organism proliferates under both anaerobic and aerobic conditions but it is preferentially micro-aerophilic in the presence or absence of combined nitrogen in the medium. It si suitable for Millets, oilseeds, fruits & vegetables, sugarcane, banana, coconut, oil palm, cotton, chilly, lime, coffee & tea, areca nut & rubber, flower, spices, & condiments, herbs, lawns & ornaments, trees etc.

Phosphobacteria

Phosphobacteria is one of the biofertilizers. Phosphorus is a major nutrient for plants inducing vigorous growth and also contributing to their disease resistance. Phosphorous helps in root formation and plant growth. The plants utilize only 10–15% of phosphate applied. The balance 85 – 90% remains in insoluble form in the soil. The phosphate solubilizing bacteria (*Bacillus megaterium*) grow and secrete organic acids, which dissolve this unavailable phosphate into soluble form and make it available to the plants. Thus, the residual phosphate fertilizers in the soil can be well utilized and external application can be optimized.

Advantages:

a) Act as a biopromoter, which facilitates root formation and plant growth.

b) It improves soil quality with subsequent uses.

Potash Mobilizing Bacteria

Potash Mobilizing Bacteria (*Frateuria aurentia*)-Use of potash-bacteria releases some of the gorged potassium between the layers of clay. The potash mobilizing bacteria is gram negative rod type bacteria, can grow in Ph 3.5 to11 and capable of mobilizing the mineral potash to the tune of 40-60 kg per ha.

Silicate Solubilizing Bacteria (SSB)

Microorganisms are capable of degrading silicates and aluminum silicates. During the metabolism of microbes several organic acids are produced and these have a dual role in silicate weathering. They supply H^{+} ions to the medium and promote hydrolysis and the organic acids like citric, oxalic acid, Keto acids and hydroxy carbolic acids which from complexes with cations, promote their removal and retention in the medium in a dissolved state.

The studies conducted with a Bacillus sp. isolated from the soil of granite crusher yard showed that the bacterium is capable of dissolving several silicate minerals under *in vitro* condition. The examination of anthrpogenic materials like cement, agro inputs like super phosphate and rock phosphate exhibited silicate solubilizing bacteria to a varying degree. The bacterial isolates made from different locations had varying degree of silicate solubilizing potential. Soil inoculation studies with selected isolate with red soil, clay soil, sand and hilly soil showed that the organisms multiplied in all types of soil and released more of silica and the available silica increased in soil and water.

Plant Growth Promoting Rhizobacteria (PGPR)

The group of bacteria that colonize roots or rhizosphere soil and beneficial to crops are referred to as plant growth promoting rhizobacteria (PGPR). The

PGPR inoculants currently commercialized that seem to promote growth through at least one mechanism; suppression of plant disease (termed Bioprotectants), improved nutrient acquisition (termed Biofertilizers), or phytohormone production (termed Biostimulants). Species of *Pseudomonas* and *Bacillus* can produce as yet not well characterized phytohormones or growth regulators that cause crops to have greater amounts of fine roots which have the effect of increasing the absorptive surface of plant roots for uptake of water and nutrients. These PGPR are referred to as Biostimulants and the phytohormones they produce include indole-acetic acid, cytokinins, gibberellins and inhibitors of ethylene production.

Despite of promising results, biofertilizers has not got widespread application in agriculture mainly because of the variable response of plant species or genotypes to inoculation depending on the bacterial strain used. Differential rhizosphere effect of crops in harboring a target strain or even the modulation of the bacterial nitrogen fixing and phosphate solubilizing capacity by specific root exudates may account for the observed differences. On the other hand, good competitive ability and high saprophytic competence are the major factors determining the success of a bacterial strain as an inoculants.

Blue-green Algae/Cyanobacteria

Blue-green algae are considered the simplest, living autotrophic plants, i.e. organisms capable of building up food materials from inorganic matter. They are microscopic. Blue-green algae are widely distributed in the aquatic environment. Some of them are responsible for water blooms in stagnant water. They adapt to extreme weather conditions and are found in snow and in hot springs, where the water is 85°C. Certain blue-green algae live intimately with other organisms in a symbiotic relationship. Some are associated with the fungi in form of lichens. The ability of blue-green algae to photosynthesize food and fix atmospheric nitrogen accounts for their symbiotic associations and also for their presence in paddy fields.

Many free-living blue-green algae (cyanobacteria) fix atmospheric nitrogen and since they are photosynthetic, they don't compete neither with crop plants nor with heterotrophic soil microflora for carbon and energy. Nitrogen-fixing ability has not only been shown by heterocystous cyanobacteria (Nostoc, Anabaena, Aulosira, etc.) but also by several non-heterocystous unicellular (Gloeocapsa, Aphanothece, Gloeothece, etc.) and filamentous (Oscillatoria, Plectonema, etc.) cyanobacteria. In nonheterocystous forms the oxygenic photosynthesis is found to be separated from nitrogen fixation either temporally or spatially. In temporal separation, nitrogen fixation predominantly occurs during the dark period and photosynthesis during the light; in these forms in terms of energy the anaerobic dark conditions are not very favorable for the process of nitrogen fixation. In spatial separation, the central non-photosynthetic cells get engaged in nitrogen fixation, whereas, the outer green cells are photosynthetically active.

The species with biofertilizer potential are the heterocystous, filamentous forms belonging to the order Nostocales and Stigonematales in which the nitrogenase activity and oxygenic photosynthesis are separated spatially and nitrogenase activity is usually light-dependent. Species of Nostoc, Anabaena, Tolypothrix, Aulosira, Cylindrospermum, Scytonema, and several other genera are widespread in Indian rice fields and contribute significantly to their fertility. Though most of the nitrogen supplementation in the soils is through the Rhizobium-legume symbiosis, the traditional use of cyanobacterial biofertilizers makes the significant contributions to rice production.

Cyanobacteria are primary colonizers and many have been shown to possess the property of tricalcium phosphate solubilization. Abundantly available rock phosphate, being insoluble, is unavailable to crop plants. Some of the cyanobacteria like, Tolypothrix, Scytonema, Hapalosiphon, etc. have been reported to solubilize rock phosphate. The cyanobacterial biofertilizer may be applied to the rice fields along with the cheaper sources of phosphorus for making phosphorus available to plants, thus allowing the sustainable use of phosphatic fertilizers. Besides the conventional approach of screening natural isolates with mineral phosphorus solubilizing ability, genetic engineering of nonphosphorus-solubilizing cyanobacteria to become efficient phosphorus solubilizer can be utilized in the cyanobacterial biofertilizers technology programme.

Azolla

Azolla is an aquatic heterosporous fern which contains an endophytic cyanobacterium, *Anabaena azollae,* in its leaf cavity. They are extremely reduced in form and specialized, looking nothing like conventional ferns but more resembling duckweed or some mosses. Azolla floats on the surface of water by means of numerous, small, closely overlapping scale-like leaves, with their roots hanging in the water. They form a symbiotic relationship with the cyanobacterium *Anabaena azollae*, which fixes atmospheric nitrogen, giving the plant access to the essential nutrient. This has led to the plant being dubbed a "super-plant", as it can readily colonize areas of freshwater, and grow at great speed - doubling its biomass every two to three days. The only known limiting factor on its growth is phosphorus, another essential mineral. An abundance of phosphorus, due for example to eutrophication or chemical runoff, often leads to Azolla blooms. The nitrogen-fixing capability of Azolla has led to Azolla being widely used as a biofertilizer, especially in parts of Southeast Asia.

Azolla mat is harvested and dried to use as green manure. There are two methods for its application in field:

a) Incorporation of Azolla in soil prior to rice plantation, and

b) Transplantation of rice followed by water draining and incorporation of Azolla.

Moreover, Azolla shows tolerance against heavy metals viz. As, Hg, Pb, Cu, Cd, Cr, etc. It tolerates low concentration but at high levels a setback in biochemical pathways is caused. *A. pinnata* absorbs heavy metals into cell walls and vacuoles through evolution of specific metal resistant enzymes. Therefore, heavy metal resistant species such as *A. pinnata* can also be incorporated as green manure in rice field near the polluted areas where heavy metal concentration is between 0.01 and 1.5 mg/liter.

Mycorrhiza as Biofertilizers

Mycorrhizas are a group of fungi that include a number of types based on the different structures formed inside or outside the root. These are specific fungi that match with a number of favorable parameters of the host plant on which it grows. This includes soil type, the presence of particular chemicals in the soil types, and other conditions. These fungi grow on the roots of these plants. In fact, seedlings that have Mycorrhiza fungi growing on their roots survive better after transplantation and grow faster. The fungal symbiont gets shelter and food from the plant which, in turn, acquires an array of benefits such as better uptake of phosphorus, salinity and drought tolerance, maintenance of water balance, and overall increase in plant growth and development.

Mycorrhizal fungi can increase the yield of a plot of land by 30%-40%. It can absorb phosphorus from the soil and pass it on to the plant. Mycorrhizal plants show higher tolerance to high soil temperatures, various soil- and root-borne pathogens, and heavy metal toxicity.

Types of Mycorrhizas

By earlier mycologists the mycorrhizas were divided into the following three groups:

a) Ectomycorrhiza. It is found among gymnosperms and angiosperms. In short roots of higher plants generally root hairs are absent. Therefore, the roots are infected by mycorrhizal fungi which, in turn, replace the root hairs (if present) and form a mantle. The hyphae grow intercellularly and develop net in cortex. Thus, a bridge is established between the soil and root through the mycelia.

b) Endomycorrhiza. The morphology of endomycorrhizal roots, after infection and establishment, remain unchanged. Root hairs develop in a normal way. The fungi are present on root surface individually. They also penetrate the cortical cells and get established intracellulary by secreting extracellular enzymes. Endomycorrhizas are found in all groups of plant kingdom.

c) Ectendomycorrhiza. In the roots of some of the gymnosperms and angiosperms, ectotrophic fungal infection occurs. Hyphae are established intracellularly in cortical cells. Thus, symbiotic relation develops similar to ecto- and endo-mycorrhizas.

The basidiospores, chopped sporocarp, sclerotia, pure mycelia culture, fragmented mycorrhizal roots or soil from mycorrhizosphere region can be used as inoculum. The inoculum is mixed with nursery soils and seeds are sown.

Benefits from Mycorrhizas to Plants

a) They increase the longevity of feeder roots, surface area of roots by forming mantle and spreading mycelia into soil and, in turn, the rate of absorption of major and minor nutrients from soil resulting in enhanced plant growth.

b) They play a key role for selective absorption of immobile (P, Zn and Cu) and mobile (S, Ca, K, Fe, Mn, Cl, Br, and N) elements to plants. These are available to plants in fewer amounts.

c) Some of the trees like pines cannot grow in new areas unless soil has Mycorrhizal inocula because of limited or coarse root hairs.

d) Vesicular Arbuscular (VA) Mycorrhizal fungi enhance water uptake in plants,

e) VA Mycorrhizal fungi reduce plant response to soil stress such as high salt levels, toxicity associated with heavy metals, mine spoils, drought and minor element (e.g. Mn) imbalance.

f) Mycorrhizal fungi decrease transplant socks to seedlings. They produce organic 'glues' which bind soil particles into semi stable in aggregates. Thus, they play a significant role in augmenting soil fertility and plant nutrition.

g) Some of them produce metabolites which change the ability of plants to induce roots from woody plant cuttings and increase root development during vegetative propagation.

h) They increase resistance in plants and with their presence reduce the effects of pathogens and pests on plant health.

Plants have a number of relationships with fungi, bacteria, and algae, the most common of which are with Mycorrhiza, rhizobium, and cyanophyceae. These are known to deliver a number of benefits including plant nutrition, disease resistance, and tolerance to adverse soil and climatic conditions. These techniques have proved to be successful biofertilizers that form a health relationship with the roots.

CONSTRAINTS IN BIOFERTILIZER TECHNOLOGY

Though the biofertilizer technology is a low cost, eco-friendly technology, several constraints limit the application or implementation of the technology the constraints may be environmental, technological, infrastructural, financial, human resources, unawareness, quality, marketing, etc. The different constraints in one way or other affecting the technique at production, or marketing or usage are as follows:

Technological Constraints

a) Use of improper, less efficient strains for production.

b) Lack of qualified technical personnel in production units.

c) Unavailability of good quality carrier material or use of different carrier materials by different producers without knowing the quality of the materials.

d) Production of poor quality inoculants without understanding the basic microbiological techniques

e) Short shelf life of inoculants.

Infrastructural Constraints

a) Non-availability of suitable facilities for production

b) Lack of essential equipments, power supply, etc.

c) Space availability for laboratory, production, storage, etc.

d) Lack of facility for cold storage of inoculant packets

e) Financial constraints

f) Non-availability of sufficient funds and problems in getting bank loans

g) Less return by sale of products in smaller production units.

Environmental Constraints

a) Seasonal demand for biofertilizers

b) Simultaneous cropping operations and short span of sowing/planting in a particular locality

c) Soil characteristics like salinity, acidity, drought, water logging, etc.

Human Resources and Quality Constraints

a) Lack of technically qualified staff in the production units.

b) Lack of suitable training on the production techniques.

c) Ignorance on the quality of the product by the manufacturer.

d) Non-availability of quality specifications and quick quality control methods.

e) No regulation or act on the quality of the products.

f) Awareness on the technology.

g) Unawareness on the benefits of the technology.

h) Problem in the adoption of the technology by the farmers due to different methods of inoculation.

i) No visual difference in the crop growth immediately as that of inorganic fertilizers.

Proper awareness and support from government and farmers will help to solve such problems. Thus, biofertilizers are important if we are to ensure a healthy future for the generations to come. Biofertilizers enhance the nutrient availability to crop plants (by processes like fixing atmosphere Nitrogen or dissolving Phosphorus present in the soil); and also impart better health to plants and soil thereby enhancing crop yields in a moderate way. It is a natural method without any problems like salinity and alkalinity, soil erosion etc. In the vast areas of low input agriculture and oil seeds production, as also in crops like sugarcane, etc, these products will be of much use to give sustainability to production. In view of the priority for the promotion of organic farming and reduction of chemical residues in the environment, special focus has to be given for the production of biofertilizers.

REFERENCES

AraujoY, Luizao F and Barros E 2004 Effect of earthworm addition on soil nitrogen availability, microbial biomass and litter decomposition in mesocosms. Biology and Fertility of Soils 39(3): 146-152.

Aseri G K, Jain N, Panwar J, Rao AV, Meghwal P R 2008 Bio-fertilizer s improve plant growth, fruit yield, nutrition, metabolism and rhizosphere enzyme activities of Pomegranate (Punica granatum L.) in Indian Thar Desert. Scientia Horticulturae 117(2): 130-135.

Berc J, Muñiz O and Calero B 2004 Vermiculture offers a new agricultural paradigm. Biocycle 45(6): 56-57. Bocchi S and Malgioglio A 2010 Azolla-Anabaena as a bio-fertilizer for rice paddy fields in the Po valley, a temperate rice area in northern Italy. International Journal of Agronomy. Volume 2010. 1-5. Available online at http://www.hindawi.com/journals/ija/2010/152158/

Celis J, Sandoval M, Zagal E y Briones M 2006 Effect of sewage sludge and salmon wastes applied to a Patagonian soil on lettuce (Lactuca Sativa L.) germination. Journal of Soil Science and Plant Nutrition 6(3): 13-25.

Chen J H 2008 The combined use of chemical and organic fertilizers and/or bio-fertilizer for crop growth and soil fertility. International workshop on sustained management of the soil-rhizosphere system for efficient crop production and fertilizer use. Taiwan: National Chung Hsing University

Chhotu J and Fulekar M 2008 Vermicomposting of vegetal waste: a bio physicochemical process based on hydro-operating reactor. African Journal of Biotechnology 7(20): 3723-3730.

Christry M and Ramaligam R 2005 Vermicomposting of sago industrial solid waste using epigeic earthworm Eudrilus eugeniae and macronutrients analysis of vermicompost. Asian Journal of Microbiology Biotechnology and Environmental Sciences 7(3): 377-381.

Classen J, Rice M and Sherman R 2007 The effects of vermicompost on field turnips and rainfall runoff. Compost Science & Utilization. 15(1): 34-39.

Coker C 2006 Environmental remediation by composting. Biocycle 47(2):18-23

Eddy D 1999 Eye on micronutrients: There are plenty of materials you can use to fertilize your crops. Here's a look at what's hot. American Vegetable Grower 47(1): 48-49.

Fresco L 2003 Fertilize the plant, not the soil. UN Chronicle 40(3): p. 62

Gharib FA, Moussa L A and Massoud O N 2008 Effect of compost and bio-fertilizer s on growth, yield and essential oil of sweet marjoram (Majorana hortensis) plant. International Journal of Agriculture & Biology 10(4): 381-387.

Jurado-Guerra P, Luna M and Barretero R 2004 Beneficial use of biosolids as organic fertilizers in arid and semiarid rangelands. Técnica Pecuaria en México 42(3): 379-395.

Kowalchuk G A, Naoumenko Z S, Derikx J L, Felske A, Stephen J R and Arkhipchenko I A 1999 Molecular analysis of ammonia-oxidizing bacteria of the β-subdivision of the class proteobacteria in compost and composted materials. Applied and Environmental Microbiology 65(2): 396-403.

Kupper K C, Bettiol W, de Goes A, de Souza P S, Bellotte J A M 2006 Bio-fertilizer for control of Guignardia citricarpa, the causal agent of citrus black spot. Crop Protection 25(6): 569-573.

Martín G, Costa Rouws J, Urquiaga S and Rivera R A 2007 Crop rotation of Canavalia ensiformis green manure of maize and arbuscular mycorrhize in an eutric rodic nitisol of Cuba. Agronomía tropical 57(4):313-321.

Mills T 2006 Composting cafeteria residuals with earthworm. Biocycle 47:54-55.

Mohamedy E and Ahmed M A 2009 Effect of bio-fertilizer s and humic acid on control of dry root rot disease and improvement yield quality of mandarin (Citrus reticulate Blanco). Research Journal of Agriculture and Biological Sciences 5(2):127-137.

Painter T J 1995 Bio-fertilizers: exceptional calcium binding affinity of a sheath proteoglycan from the blue-green soil alga Nostoc calcicola. Carbohydrate Polymers 26(3): 231-233.

Peigné J and Girardin P 2004 Environmental impacts of farm-scale composting practices. Water Air Soil Pollution 153(1-4): 45-68.

Petiot C and Guardia A 2004 Composting in a Laboratory Reactor: A review. Compost Science & Utilization 12(1): 69-79.

Rajendran K y Devaraj P 2004 Biomass and nutrient distribution and their return of Casuarina equisetifolia inoculated with bio-fertilizer s in farm land. Biomass and Bioenergy 26(3): 235 –249.

Reddy K and Shantaram M 2005 Potentiality of earthworms in composting of sugarcane byproducts. Asian Journal of Microbiology Biotechnology and Environmental Sciences 7(3): 483-487.

Rivera-Cruz M, Trujillo A, Córdova G, Kohler J, Caravaca F, Roldán A 2008 Poultry manure and banana waste are effective bio-fertilizer carriers for promoting plant growth and soil sustainability in banana crops. Soil Biology & Biochemistry 40(12):3092-3095.

Roberts P, Edwards-Jones G, and Jones D L 2007 Yield responses of wheat (Triticum aestivum) to vermicompost. Compost Science & Utilization 15(1): 6-15.

Rosset P 1998 Alternative Agriculture Works: the case of Cuba. Monthly Review 50(3):137-146.

Sessitsch A, Howieson J G, Perret X, Antoun H and Martínez-Romero E 2002 Advances in Rhizobium research. Critical Reviews in Plant Sciences 21(4): 323-328.

Singh A and Sharma S 2003 Effect of microbial inocula on mixed solid waste composting, vermicomposting and plant response. Compost Science & Utilization 11(3): 190-199.

Stamford N P, Santos P R, Santos C E S, Freitas A D S, Dias S H L, and Lira Jr M A 2007 Agronomic effectiveness of bio-fertilizer s with phosphate rock, sulphur and Acidithiobacillus for yam bean grown on a Brazilian tableland acidic soil. Bioresource Technology 98(6):1311-1318.

Suthar S 2005 Vermicomposting of kitchen waste by using Eisenia foetida (Savigny). Asian Journal of Microbiology Biotechnology and Environmental Sciences 7(3): 541-544.

Thenmozhi R, Rejina K, Madhusudhanan K and Nagasathya J 2010 Study on effectiveness of various bio-fertilizers on the growth & biomass production of selected vegetables. Research Journal of Agriculture and Biological Sciences 6(3): 296-301.

Tognetti C et al 2005 Composting vs. vermicomposting: A comparison of end product quality. Compost Science & Utilization 3(1): 6-13.

Welbaum G, Sturz A, Dong Z and Nowak J 2004 Managing soil microorganisms to improve productivity of Agro-ecosystems. Critical Reviews in Plant Sciences 23(2): 175-193.

Index